MUTANTS, ANDROIDS, AND ALIENS

On Being Human in the Marvel Cinematic Universe

James A. Tyner

University Press of Mississippi / Jackson

The University Press of Mississippi is the scholarly publishing agency of the Mississippi Institutions of Higher Learning: Alcorn State University, Delta State University, Jackson State University, Mississippi State University, Mississippi University for Women, Mississippi Valley State University, University of Mississippi, and University of Southern Mississippi.

www.upress.state.ms.us

The University Press of Mississippi is a member of the Association of University Presses.

Any discriminatory or derogatory language or hate speech regarding race, ethnicity, religion, sex, gender, class, national origin, age, or disability that has been retained or appears in elided form is in no way an endorsement of the use of such language outside a scholarly context.

Copyright © 2025 by University Press of Mississippi
All rights reserved
Manufactured in the United States of America

∞

Publisher: University Press of Mississippi, Jackson, USA
Authorised GPSR Safety Representative: Easy Access System Europe - Mustamäe tee 50, 10621 Tallinn, Estonia, gpsr.requests@easproject.com

Library of Congress Control Number: 2025004973

Hardback ISBN: 9781496857385 | Trade Paperback ISBN: 9781496857392
Epub Single ISBN: 9781496857408 | Epub Institutional ISBN: 9781496857415
PDF Single: 9781496857422 | PDF Institutional ISBN: 9781496857378

British Library Cataloging-in-Publication Data available

To Stuart Aitken

CONTENTS

Preface. ix
Acknowledgments xix

Chapter One: Humans, Assemble! 3

Chapter Two: Marvel's Monstrous Other 27

Chapter Three: Marvel's Modern Prometheus 64

Chapter Four: Marvel's Alien Encounters 95

Chapter Five: Endgames. 121

Notes 131
Bibliography 167
Index 185

PREFACE

Stealing bodies, talking about aliens, and bringing the dead back to life.
What kind of creature feature did I sign up for?
—ALPHONSO "MACK" MACKENZIE (*AGENTS OF S.H.I.E.L.D.*,
SEASON 2, EPISODE 7, "THE WRITING ON THE WALL")

During the course of writing this manuscript, I have come to empathize with Alphonso "Mack" Mackenzie.[1] As an agent of S.H.I.E.L.D., Mack is tasked with defending humanity against all manner of "creatures," including enhanced humans, androids, and extraterrestrials. Mack is not a superhero—or at least, not as this identity is conventionally understood. Mack is human and he lacks superpowers. He is not a cyborg and, unlike Iron Man or War Machine, he does not wear an exoskeleton. To be sure, Mack is capable of performing extraordinary feats, especially in the face of danger and possible death. He is, after all, a trained agent. Mack's character is important, however, precisely because he is a rather *ordinary* human being. As such, he struggles with the ethical challenges of living among—and potentially killing or being killed by—myriad other-than-human and more-than-human beings.

In both literature and cinema, mutants, androids, and aliens have long functioned as humanity's Other, with their "nonhuman" bodies serving as surrogates to explore humanity's prejudices, bigotry, and hatred.[2] To date, scholars working in several fields, including feminism, ethnic studies, critical race studies, Queer studies, and disability studies—to name but a few—have deconstructed the representations of the Othered body and how fictional depictions provide insight into the contested terrains of identity, subjectivity, and personhood. In science fiction more broadly, and the superhero genre in particular, the fictional Other often exists as "example against which the "normal" human is juxtaposed."[3] Most commonly, however, the focus is directed toward the superhero as Other. In other words, it is the superhero's body, or perhaps that of the supervillain, that is under investigation. As a result, the "regular" or "ordinary" humans remain in the background. Charting a parallel path, I suggest that it is important to take seriously the "ordinary" humans who ally with or oppose superheroes: the law enforcement officers, military officials, politicians, and countless other nameless civilians who try to make sense of a rapidly changing more-than-human and other-than-human universe. To that end, in *Mutants, Androids, and Aliens*, I provide a critical posthumanist reading of being human in the Marvel Cinematic Universe (MCU).

Posthumanism is a concept imbued with multiple meanings, not all of which are consistent across and within disciplines.[4] Indeed, posthumanism is an overburdened term, a concept at risk of collapsing under the weight of academic largess. For my present purposes, posthumanism refers to the sense that our traditional view of what constitutes a human being is undergoing a profound transformation.[5] This introspective moment derives from revolutionary advances largely in the field of biotechnology. Innovations in robotics, prosthetics, machine intelligence, nanotechnology, and genetic manipulation—in short, the "stuff" of the superhero genre—are changing how *we* understand ourselves and our relations with other, nonhuman and more-than-human beings.[6] Accordingly, posthumanism is less about a world populated by cyborgs and androids and artificial intelligence systems, and more about the relationships that comprise everyday life among the myriad beings that exist *and may yet come into existence.*

Posthumanism challenges and critiques long-held assumptions that presume humans as the center of meaning, value, knowledge, and action.[7] In so doing, posthumanism presents a radical critique of being human in a "more-than-human" or "other-than-human' world."[8] Posthumanism rejects the view of the human as exceptional, as a being separate from, and (usually) dominant over, other life forms.[9] That is, posthumanism critiques the deep-seated belief that humans—however embodied—are endowed with unique qualities and properties that allow them to hold dominion over other-than-human beings. In its rejection of the human as exceptional, as a being separate from all others, posthumanism constitutes a refusal to posit that only humans are the "subject-of-a-life" and thus preeminent in their relations with other forms of being. Stated differently, this rejection of an anthropocentric human exceptionalism is an affirmation of life beyond the human. To think against human exceptionalism, as James Bridle concludes, is to override our human tendency to separate ourselves from the natural—and the so-called artificial—world.[10]

Posthumanism is not nihilistic; it does not call for the negation of the human or for the eradication of humanity.[11] Our anticipated (or feared) more-than-human and other-than-human futures will still include "ordinary" humans, with all their faults and frailties. Barring an extinction-level event, nonenhanced humans will remain on the scene. And for many (most?) of us, the heralded benefits of biotechnologies and genetic engineering will remain chimeric. If Mars or some other distant celestial body is colonized—perhaps because life on Earth has become near-impossible—the lion's share of humanity will remain firmly if not fearfully rooted in its ruined place.[12] The principal challenge that posthumanism poses is thus to reposition the human, that is, to "track and analyze the shifting grounds on which new, diverse, and even contradictory understandings of the human are currently being generated."[13] Posthumanism, in other words, remains interested and invested in the human experience, insofar that posthumanism seeks to understand human life as it is lived with and through its relations with those beings conventionally excluded from being fully human. As such, posthumanism reconceives being human in new ways and provides a much expanded understanding of the human.[14] These concerns are manifest throughout the MCU.

The superhero genre is vast and the corpus of superhero genre critique is familiar terrain: American exceptionalism, overt militarism, the reproduction of whiteness, masculinity, heteronormativity, disability, and hyper-nationalism. Esther De Dauw, for example, demonstrates—rightfully so—the ways in which the superhero genre supports normative ideas and ideals related to race, gender, and sexuality.[15] Markedly, much of this scholarship has criticized the genre—including the MCU—for reinforcing an exclusionary understanding of human beings and of being human. The superhero characteristically embodies the dominant archetype of the able-bodied, heterosexual white man, as represented by Thor, Steve Rogers (Captain America), Tony Stark (Iron Man), Peter Quill (Star Lord), and Dr. Strange. Echoing these concerns, Samira Nadkarni justifiably critiques the persistent portrayal of American exceptionalism and imperialism throughout the MCU. Analyzing *Agents of S.H.I.E.L.D.*, she charges that "the MCU participates in contemporary and historical propaganda that positions the United States and its allies as performing humanitarian militarism from.... a notable position of power—that of speaking for and acting in service of an oppression portion of humanity—and therefore recasting the performance of violence upon the other seemingly in service of re-humanizing an other."[16]

That said, since the mid-2010s, the Marvel franchise has made strides, or, perhaps, small steps, to diversify the MCU and to challenge racist, sexist, and ableist prejudices, as demonstrated by the appearances of T'Challa (Black Panther), Shang-Chi, Wong, Natasha Romanoff (Black Widow), Wanda Maximoff (Scarlet Witch), Carol Danvers (Captain Marvel), Monica Rambeau, Kate Bishop, Daisy Johnson (Quake), Melinda May, and Kamala Khan (Ms. Marvel), to name but a few. Similarly, Loki is revealed as bisexual and America Chavez (Ms. America) is the daughter of a same-sex couple, Elena and Amalia Chavez. Matt Murdock (Daredevil) is visually impaired and both Clint Barton (Hawkeye) and Maya Lopez (Echo) are hearing-impaired; the latter character also has a prosthetic leg. In addition, the franchise has addressed broader structural forms of discrimination, including racism, sexism, and, to a lesser extent, homophobia. In the streaming series *The Falcon and the Winter Soldier*, for example, Sam Wilson routinely confronts racism when he dons the mantle of Captain America.[17]

For other critics, the superhero genre is a superficial genre, one that has little redeeming artistic or aesthetic value.[18] Indeed, the superhero genre in general, and the MCU in particular, is the bane of many commentators, both within the film industry and in academia. The actress Rose McGowan, for example, describes superhero films as lacking in "complexity, story development, character development, and freedom of thought."[19] Most vocal in the condemnation of the superhero genre is the trenchant critique offered by director Marin Scorsese. For Scorsese, cinema is "about revelation—aesthetic, emotional, and spiritual revelation." He continues that cinema is "about characters—the complexity of people and their contradictory and sometimes paradoxical natures, the way they can hurt one another and love one another and suddenly come face to face with themselves." Cinema, in other words, constitutes "an art form" and, on this point, Scorsese concludes, the superhero genre falls short. Addressing Marvel films specifically, Scorsese is dismissive, claiming that there is no revelation, no mystery, and no genuine emotional danger; in short, "nothing is at risk."[20] The director Francis Ford Coppola is blunter, describing Marvel movies as "despicable."[21]

In subsequent chapters, I acknowledge and draw on these critiques; however, it is not my intention to reproduce these arguments. In fact, I find myself in disagreement with many of the accusations directed toward the superhero genre, notably that these films fail to capture something of the complexity of being human and all that reveals. Instead, the path I have chosen is rather unorthodox in that I direct my primary concern away from the stalwarts of the MCU—Iron Man, Captain America, Thor, and the rest—to the more mundane human characters who stand with, or against, Earth's mightiest heroes. Indeed, my decision to begin this journey with Mack, an important but secondary character in a Marvel series that is not normally considered canon in the greater MCU, is unusual in and of itself.[22]

Matthew Ellis, Ellen Nadeer, Rosalind Price: These human characters are unfamiliar to all but the most die-hard of fans. This is hardly surprising—audiences purchase theater tickets or pay for streaming services to watch superheroes battle supervillains: Captain America battles Red Skull, Spider-Man fights the Green Goblin. The unfortunate consequence, as Daniel Drezner observes, is the steady disappearance of ordinary civilians from the narrative.[23] And yet, ordinary humans do feature in many of the shows and their on-screen presence serves important plot devices, for the MCU diegetic often turns on whether humans show compassion to enhanced individuals—like Erik Selvig does—or scapegoat and persecute more-than-human and other-than-human beings—like J. Jonah Jameson does—because of fear, suspicion, and hatred. Therefore, instead of focusing my attention exclusively, or even primarily, on superheroes and supervillains, I reconsider those ordinary humans who coexist alongside superheroes and supervillains. *Mutants, Androids, and Aliens*, accordingly, keeps humans in the spotlight. This requires a re-centering of Marvel's secondary human characters, including the aforementioned Ellis, Nadeer, and Price but also Jimmy Woo, Sadie Deever, Holden Radcliffe, and many others. To state the obvious: Not all beings who inhabit the MCU are superheroes and it is time to reassemble the humans alongside their more famous other-than-human and more-than-human counterparts.

That said, my purpose is decidedly *not* to reclaim humanity's supremacy as a species, but instead to provide a searching exploration of *what it means to be human* in Marvel's expansive cinematic universe. My focus is less on the physical or psychological attributes of individual superheroes and supervillains and more on the social relations between them and their human counterparts. My endgame is to say something about the meaningfulness of life itself and of the necessity to find empathy and compassion in a violent world filled with other-than-human and more-than-human beings. Moving forward, I want to maintain the focus on the human characters in the MCU, to consider more concretely how these characters reflect on and respond to living among more-than-human and other-than-human beings.

This is a subtle but not insignificant shift, in that I am less concerned with how "we"—the viewing audience, scholars, and commentators—(re)interpret mutants, androids, and aliens, and more focused on how fictional human beings perceive mutants, androids, and aliens and thus experience being human in a posthuman universe. Certainly, my approach is not without its own set of challenges. To conceive of mutants, androids, and aliens as the irretrievably but metaphorical monstrous Other is in itself to adopt an anthropocentric perspective—a strategy seemingly discordant with much posthumanist thought.[24] Ultimately, though, I believe such an approach is important if only to offer a counterpoint to the more dominant critiques currently available. Catherine Marshall, for example, asks if humans will ever collectively be able to access the humanity of the stranger—especially the stranger who is so different from us that we are certain we will never succeed in finding common ground?[25] A posthumanist critique of being human in Marvel's cinematic universe provides a possible path toward resolving this question, insofar as posthumanism calls for a more inclusive moral-ethical response and responsibility to nonhuman life forms and, indeed, toward other beings not necessarily thought of as being alive.[26] Consequently, a simple question guides my project is: How might humans adapt meaningfully and morally to the existence of more-than-human and other-than-human beings?[27]

Moral standing, following Martin Schönfeld, "is a normative extension of traditional moral discourse with the aim to incorporate parts or the whole of nature into the domain of moral relevance." However, as Schönfeld clarifies, "finding out who, or what, has moral standing among non-human beings requires investigating just how far the quality of moral standing can be justifiably extended to parts or the whole of nature."[28] Do we, for example, extend moral standing to all life or perhaps only to other sentient beings?[29] And what are we to make of extraterrestrial beings, should these be found to exist? Similarly, should moral standing be extended to artificial intelligence systems? Perhaps machines, notably robots and androids, should be included in humanity's moral community and not, as some people suggest, be seen purely as instruments designed and produced to satisfy human needs.

Superheroes and supervillains, as demonstrated by their missions, often possess strict and clearly defined moral codes.[30] Captain America, for example, is understood both by his fictional counterparts and by real-world audiences as the moral compass of the MCU. The High Evolutionary, who makes his on-screen debut in *Guardians of the Galaxy Vol. 3*, is a cruel vivisectionist wholly lacking of compassion or empathy. That said, Marvel has often subverted the moral clarity of its characters. Indeed, a distinctive attribute of most characters in the MCU is that of moral ambiguity. Setting aside perhaps Steve Rogers, all superheroes are fundamentally flawed: Thor is a rash and impulsive god; Tony Stark is an egotistical philanderer; and Bruce Banner with his internal doppelganger, The Hulk, is the ultimate embodiment of the dualism of good and bad. Supervillains, likewise, are rarely figures of pure malice; rather, their turn to the dark side often is the result of a series of tragic events. In the end, though, while a character's morality is not necessarily sacrosanct, audiences generally have no problem discerning good from evil. Stark is and always will be a hero; Thanos is irredeemably a villain.[31]

But what of the many ordinary humans who inhabit an increasingly more-than-human and other-than-human universe? An under-theorized diegetic of Marvel's speculative universe is the ways in which humans do or do not extend moral standing to more-than-human and other-than-human beings. In other words, what are the moral codes of those humans who aren't superheroes or supervillains? Do humans extend moral standing to more-than-human and other-than-human beings or are these "nonhumans" forever excluded from humanity's restricted moral community? As a foray into Marvel's cinematic portrayal of a posthuman future, this book explores the possible ways humans *might* extend moral standing to other-than-human and more-than-human beings. As Nash describes, many people find compelling the notion that nonhuman life and nonliving matter have (or should have) moral standing.[32] I share this sentiment. I agree with those who critique and challenge the anthropocentric dogma that humans are the normative standard of adjudicating who—or what—is accorded moral worth. I take seriously the ethical position that moral standing should be extended to more-than-human and other-than-human beings. And to that end, I frame my study on a form of biocentrism proposed by the theologian and philosopher Albert Schweitzer.[33]

Schweitzer's ethics are expressed in a simple statement: "It is good to maintain and to encourage life; it is bad to destroy life or to obstruct it."[34] This deceptively straightforward proposition, I argue, provides a foundation through which we can better appreciate the meaning of being human in a posthuman universe. To begin, Schweitzer's ethics of reverence for life resonates strongly with the aforementioned moral absolutes that often define the superhero genre. For Schweitzer, evil is that which hampers or hinders life; goodness is that which saves or helps life. And yet, Schweitzer concedes that moral absolutes do not exist in reality; instead, living beings—like their fictional counterparts in the superhero realm—are confronted regularly with ethical uncertainties that often mark the difference between life and death.

Admittedly, Schweitzer was—and remains—a controversial figure in philosophy and his appearance in a book on superheroes is unconventional. His ethos that *all* life is not to be denied but affirmed has been derided as naïve, impractical, and untenable. However, following Ara Barsam, the reason we bother with Schweitzer is because we gain a better appreciation of the contribution that reverence for life makes to contemporary ethical concerns.[35] In Marvel's fictional world, we repeatedly confront ethical decisions made by more-than-human and other-than-human beings but also, just as important, choices made by everyday human beings who struggle to find their place in a rapidly changing universe.

For Schweitzer, there are two ways of looking at the world—one is optimistic and full of hope, the other is pessimistic and full of despair.[36] And for Schweitzer, only the former offers a path toward living a meaningful life and this entails an ethos of care and compassion not only for humans, but also for the panoply of more-than-human and other-than-human beings that share our universe. Simply put, for Schweitzer, the diversity and multiplicity of all beings manifest a shared inner essence: a will to live.[37] As such, a posthuman ethos centered around Schweitzer's will-to-live necessarily requires an active participation in helping all living beings—and those not normally considered to be living—to survive and thrive throughout the universe.

Schweitzer's concept of reverence for life, Mike Martin explains "is not an amorphous affirmation of living creatures that leaves us without practical guidance," but instead "provides guidance through specific ideals that function as signposts in a morally complex and ambiguous world."[38] It is a declaration that killing is always morally—if not legally—wrong. It is a covenant to always strive to do better. As such, following Schweitzer, "The will-to-live, which is filled with reverence for life, is interested in the most lively and preserving way that can be imagined."[39] This is a tall order for the superhero genre, which, almost by definition, pivots on life-and-death struggles between opposing combatants. Few if any fans would sit through a two-hour *action* film to see nothing but diplomacy and goodwill. However, the superhero genre *does* provide a unique opportunity to carry out this objective. On this point, I read the superhero genre against the grain, that is, to critically interrogate the normative ethics that inform the social relations of ordinary humans and their more-than-human and other-than-human counterparts.

And why, specifically, the MCU? Simply stated, the franchise now comprises an unprecedented series of interlinked films, television series, and "one-shot" productions that have pushed the boundaries of transmedia entertainment in general and the superhero genre in particular. Indeed, as Joanna Robison and coauthors explain, the films—connected and complicated by overlapping plotlines and dozens of television shows—have formed a vast tapestry of characters, incident, and high emotion.[40] To date, the MCU is far and away the most successful franchise within the superhero genre.[41] Financially, the thirty-plus films within the MCU to date have tallied over $28 billion worldwide.[42] Of the top ten highest grossing superhero films (as of 2021), seven belong to the MCU; of the top thirty superhero films, sixteen are from the MCU.[43] The MCU is so successful that that five of the top ten all-time grossing films are part of the franchise.[44] However, critical study of the MCU is warranted not solely because of its financial successes, but also because of the possibilities Marvel's fictional universe provides to question what it means to be human in a speculative universe populated both with more-than-human and other-than-human beings. Indeed, in Marvel's world, humans are overshadowed and—in the words of Nick Fury—"hopelessly, hilariously outgunned" by nonhuman beings: extraterrestrials, mutants, Inhumans, robots, androids, and countless other fantastical creatures.[45] Steve Rogers (Captain America), for example, is a genetically enhanced human; he received his superhuman powers after being injected with the Super Soldier

Serum created by Abraham Erskine, whereas Tony Stark (Iron Man), Sam Wilson (Falcon), and James Rhodes (War Machine), derive their powers through technological augmentation, namely militarized exoskeletons. Conversely, Bruce Banner (The Hulk), Peter Parker (Spider-Man), and Carol Danvers (Captain Marvel) have been empowered through accidents that altered their genetic composition. Daisy Johnson (Quake) and Elena "Yo-Yo" Rodriguez are Inhumans; these beings are genetically close to humans but have enhanced powers that are manifest after undergoing a process known as Terrigenesis. More "conventional" extraterrestrials include, among others, the Kree, the Chitauri, the Skrull, and the Asgardians, including Thor. Although described frequently as being a god, Thor (similar to all other Asgardians) is not immortal. Peter Quill (Star Lord), however, is a being born of a human mother and a God-like being (strictly speaking, a Celestial). Cyborgs, robots, and androids make frequent appearances in the MCU. Here, distinctions matter. Robots are fully mechanical artificial devices with no organic elements; androids are robots that resemble human beings. The supervillain Ultron is a robot, whereas AIDA is a type of android known as a "Life Model Decoy." Unlike robots and androids, cyborgs are techno-organic hybrids, that is, humans with "artificial" devices. These technological modifications can be implanted, as with cochlear implants, deep brain stimulators, and cardiac pacemakers; also included are external modifications, such as prosthetics. The character Razer Fist, appearing in *Shang-Chi and the Legend of the Ten Rings* (2021), is a cyborg, in that he is a human with a steel blade for a hand. Mike Peterson (Deathlok), a human who appears in *Agents of S.H.I.E.L.D.*, is likewise a cyborg; his transformation is detailed throughout several episodes of the series. Strictly speaking, however, Mike Peterson is a hybrid cyborg, in that he initially received enhanced powers after being injected with the Extremis drug—similar to the process undertaken by Steve Rogers; Peterson is later transformed into a cyborg by a sinister organization known as HYDRA. Indeed, the character of Mike Peterson highlights the fluidity of "being" in the MCU. Tony Stark, technically, is also a cyborg, given that he stays alive with the assistance of an artificial heart device. Bucky Barnes (Winter Soldier) is equally complex; born human, he was subsequently injected with the Super Soldier Serum and outfitted with a weaponized prosthetic arm. Lastly, there are humans who acquire "superhero" powers through extensive training; included here are Clint Barton (Hawkeye), Natasha Romanoff (Black Widow), Melinda May, and Stephen Strange (Dr. Strange). Faced with such a panoply of more-than-human and other-than-human beings that blur the distinction of life itself, it is hardly surprising that humanity in Marvel's cinematic universe is pressed to turn inward, to contemplate anew its place in the world. It is our task to contemplate this new relation.

ACKNOWLEDGMENTS

Writing a book on superheroes is nerve-racking. Personally, the angst doesn't come from the actual writing; instead, for me, it stems from a hope to reach two sometimes disparate audiences. On the one hand, there are the fans. For these readers, attention will focus often on the details of the characters and their fictional worlds: of the nuances of plot development, continuity, and cohesion. And—let's be honest—fans enjoy reading about their favorite characters and may be disappointed if those characters fail to make an appearance. On the other hand, there are academics. Certainly, academics can be fans of superheroes, and many scholars I know do enjoy and appreciate the genre. That said, many academics cast a critical eye toward superheroes, and frequently not without reason. So where does this leave me? My objective has been always to approach the superhero genre as both a fan and an academic; thus, it is to see the (fictional) characters as subjects who give meaning to (real) life.

Many thanks to Emily Bandy, Joey Brown, Katie Turner, Laura Strong, and the entire staff at University Press of Mississippi for their professionalism in taking this project from its early conception to the final product. Heartfelt thanks are extended to Mary Andino, for her thorough copyediting of the manuscripts; any errors that remain are mine alone. Thanks also to those scholars who reviewed the initial prospectus and to those who read the completed manuscript. At every stage, countless individuals, some known, many anonymous, provided critical feedback, constructive criticism, and invaluable support.

I thank my family for their encouragement: first and foremost, my parents, Dr. Gerald Tyner and Dr. Judith Tyner, for their inspiration and encouragement over the years, for the sacrifices they made as I pursued my dreams. I thank also my daughters, Jessica and Anica Lyn; and especially my partner in life, Belinda, who is my superheroine. Belinda: Your empathy is boundless, and I am forever in your debt. I must acknowledge also Carter and Bubba, our eight-year-old and five-year-old rescue dog and cat.

And finally, I thank Stuart Aitken. Many years ago, I worked as a teaching assistant for Stuart in the geography program at San Diego State University. Stuart was going to attend a conference and asked me to give a lecture to his Introduction to Cultural Geography class. Since that first opportunity to share my ideas to an audience, Stuart has supported my career and has become a friend. It is in the spirit of his kindness and generosity that I warmly dedicate this book to Stuart.

MUTANTS, ANDROIDS, AND ALIENS

Chapter One

HUMANS, ASSEMBLE!

You know we're prehistoric creatures oozing out of the swamps,
through the mist, clawing our way onto land so we can finally stand. Or
plant roots. Or fly.
—S.H.I.E.L.D. AGENT JOHN GARRETT (*AGENTS OF S.H.I.E.L.D.*,
SEASON 1, EPISODE 22, "BEGINNING OF THE END")

In *Avengers: Endgame*, Tony Stark faces a grave decision. It's been five years since Thanos used the Infinity Stones to eradicate half of all life in the universe, including half of humanity. With the possibility of time travel, however, Stark and the Avengers have an opportunity to undo the deaths caused by Thanos. Stark, though, has finally found meaning in life. During the intervening years—during the so-called "Blip"—Stark marries his long-time partner, Pepper Potts and they have a daughter, Morgan. Late one evening, Stark and Potts reflect on his decision to travel back in time. "We got really lucky," Potts concedes, "A lot of people didn't." "I can't help everybody," Stark responds. Grimly, Potts answers: "It sort of seems like you can."[1] Later, a reluctant Stark confronts Steve Rogers. "We've got a shot at the Stones. I just need you to know my priorities. Bring back everyone we lost, hopefully. Keep what I found, definitely. And, let's not die trying." In the end, Stark does die, sacrificing his life to restore those who were killed by Thanos.

During Stark's funeral, a holographic recording he prepared is played for the audience. "Everybody wants a happy ending," the message begins, "but it doesn't always roll that way." The recording continues:

Maybe this time. I'm hoping if you play this back, it's in celebration. I hope families are reunited. I hope we get it back and something like

Tony Stark (Robert Downey Jr.) confronts his own mortality in *Avengers: Endgame* (2019).

a normal version of the planet has been restored. If there ever was such a thing. God, what a world. Universe, now. If you told me ten years ago that we weren't alone, let alone, you know, to this extent, I mean, I wouldn't have been surprised, but come on. The epic forces of darkness and light that have come in to play. And for better or worse, that's the reality Morgan's gonna have to find a way to grow up in. So I thought I better record a little greeting, in the case of an untimely death, on my part. I mean, not that death at any time isn't untimely. This time travel thing we're gonna try and pull off tomorrow, it's got me scratching my head about the survivability of it all. Then again, that's the hero gig. Part of the journey is the end. What am I even tripping for? Everything's gonna work out exactly the way it's supposed to.[2]

There is a misperception that "moral absolutes" exist in the superhero world, that a clear boundary exists between right and wrong, good and evil.[3] Certainly, some characters are depicted in unambiguous ways. Captain America fights for those who are bullied and Red Skull covets global domination. Terence McSweeney, on this point, writes of the "illusory moral ambiguities" of the superhero. He contends that the supposedly "subversive" nature of the MCU fails to deliver "in anything more than superficial ways." Instead, the MCU reaffirms a particular morality that continues to promote nothing more than "American foundational mythic values" as embodied especially by Captain America.[4] Simply put, McSweeney concludes, "This is how the MCU views the world: with the United States as a reluctant operator in global events, but one that is necessary and entirely moral in its actions."[5]

McSweeney perhaps overstates his argument, for it is not always the case that clear boundaries can be drawn around most characters. Such moral ambiguity, in fact, is what separates Marvel's fictional world from that of other superhero worlds. Since Marvel comics first appeared on the scene, their heroes have traditionally been outsiders, shunned by society for being different.[6] Even the villains are often cast in a sympathetic light.[7] Indeed, the MCU is popular among many fans precisely because the characters are flawed and the lines between "good" and "evil" and "right" and "wrong" are blurred. As the bio-fabricated being Vision quips in *Avengers: Age of Ultron*, "Maybe I am a monster. I don't think I'd know if I were one. I'm not what you are and not what you intended."[8]

Throughout the MCU, both human and other-than-human characters are deeply damaged—emotionally, psychologically—and rarely act with unwavering conviction. Tony Stark is a particularly flawed character and has been subjected to considerable critique among scholars.[9] Ashley Robinson, for example, finds that Stark "embodies an America characterized by unchecked capitalism, corporate greed, narcissism, stagnation, and a powerful military-industrial complex."[10] And while Stark "grows as a character," Joseph Zornado and Sara Reilly concede, "from a selfish billionaire boozehound to selfless, self-sacrificing savior who dies so that all can live," he remains an icon of a conservative American national identity.[11] Indeed, according to Zornado and Reilly, Stark's "death is the necessary messianic ending" that both saves the world and protects the status quo from which he benefitted in life.[12]

My purpose here is not to adjudicate the morality of Stark or of any other character in the MCU. Both fans and detractors of the Avengers have and will continue to debate incessantly on the righteousness (or not) of Iron Man, Captain America, Nick Fury, and the other inhabitants of Marvel's cinematic universe. In the end, these are fictional characters and remain subject to multiple (and contradictory) interpretations. Instead, I maintain that life and death in the world of the superhero provides an opportunity to better understand our being human in a posthuman universe. This in turn allows for a deeper reflection on how human and other-than-human and more-than-human characters make ethical decisions within an inherently dangerous, violent universe.

Certain philosophical questions, Russell Blackford explains, recur in stories that involve mutants, androids, and aliens; paramount among these questions is whether—or at what point—we should count these beings as human or as morally equivalent to being human. That said, Blackford underscores that a pressing question as we head deeper into a posthuman future is how we (humans) are to live and to construct societies in a world inhabited by more-than-human and other-than-human beings. In other words, it is necessary to consider the interrelations of species within a posthuman framework and this, in turn, requires a consideration of humanity's reverence for more-than-human and other-than-human life.

Ethics in the MCU are always and already relational and this suggests that our moral framing needs to refocus attention away from superheroes and supervillains acting in isolation and instead toward their interactions with "ordinary" humans. Ultimately, the speculative world of Marvel beckons not a definitive judgement, but a conjecture of how a future humanity might struggle with finding meaning in a universe teeming both with "biological" and "artificial" other-than-human beings. To that end, it is helpful to remember that many ethical questions center on attitudes toward life itself. Does humanity recognize other-than-human and more-than-human beings as having intrinsic worth or, conversely, as instrumental and utilitarian? Will humans respond with care and compassion or cruelty and violence? Will humans hunt the Other, or, instead, work to extend moral standing to nonhuman beings? Before addressing these moral concerns, however, it is necessary to rethink humanity's own origin story, of how *we* became human and, in turn, struggle to find meaning in being human.

BECOMING HUMAN

All life on Earth is carbon-based. Not this guy. Whatever he runs on, it's not on the periodic table.
—UNNAMED MEDICAL EXAMINER (*CAPTAIN MARVEL* [2019])

You're saying he's not from around here?
—NICK FURY (*CAPTAIN MARVEL* [2019])

Who are we? *What* are we? Such questions speak to existential questions that have long shaped humanity's self-perception and are reflected in our myths and religions but also our sciences. Within the superhero genre, Carnes and Goren explain, "origin stories are foundings" and "they shape our understanding of the character, and they also provide important context by describing the community in which the hero exists."[13] Taking this

In *Agents of S.H.I.E.L.D.*, the "Absorbing Man," Carl Creel (Brian Patrick Wade) has the power to absorb any material that he touches.

cue, in the following two sections I provide a brief origin story both of the biological roots of becoming human and, subsequently, of the philosophical quest for the meaning of being human. This is needed, I maintain, to better understand how superheroes and supervillains "exist" within our reality, our own universe, as we have come to know it. So conceived, it thus becomes possible to reflect on the morals and ethical choices humans exhibit in a rapidly diversifying posthuman universe.

Let's begin with a fairly banal statement: Human beings are material beings. Indeed, every corporeal being that exists is made up of matter: rocks, trees, flowers, dogs, whales, and walruses. Matter, here, simply refers to anything that has both mass and volume.[14] Around 420 BCE, the Greek philosopher Empedocles's proposed that all matter was derived from four primordial substances: earth, fire, water, and air. Empedocles' thesis contributed to the fictional creation by Marvel Comics of the superhero team, the Fantastic Four. Conceived in 1961 by Stan Lee and Jack Kirby, the mutated humans who comprise the Fantastic Four included Mister Fantastic, whose plastic body resembles the fluidity of water; the Invisible Woman, who can render herself indiscernible like air; the Human Torch, who can conjure fire at will; and the Thing, made up of rocks and possessing superior strength. In time, other elemental characters were introduced, notably Iron Man and the Silver Surfer, a trend that continues in the MCU today. In *Agents of S.H.I.E.L.D.*, for example, several Inhumans are the material embodiment of basic elements, most notably Carl Creel (the "Absorbing Man") who, having been subjected to the Particle Infusion Chamber, is able to replicate the material properties of any element that he touches.

Contrary to Empedocles, all matter on Earth is composed not of four primordial elements but instead of ninety-four naturally occurring elements and twenty-four synthetic elements. How the elements and the material universe came into being remains a mystery.[15] What is commonly accepted is that our known universe was formed approximately 13.7 billion years ago following a cosmic explosion known as the Big Bang. According to this theory, the dimensions of space were compressed into a single point and time—a so-called singularity.[16] That said, there could have been an infinite number of universes constantly being formed at different times; this multiverse—of which the MCU has mined liberally—would be infinite and eternal, containing endless cycles of different universes, with no beginning in time.[17]

Still, we don't know precisely *how* the universe began. At the moment of the Big Bang, all matter was compressed into *something* smaller than an atom and neither time nor space existed. There was *nothing* else. And yet, in a fraction of a fraction of a second, as David Baker writes, "our destiny was already etched into the very fabric of the cosmos."[18] In a process termed nucleosynthesis, the "light" or "primordial" elements— were formed: hydrogen, deuterium, helium, and lithium. The heavier elements—including carbon, nitrogen, and silicon—came into existence hundreds of millions of years later, when the first stars formed. Then, in a process known as stellar nucleosynthesis, the remaining elements essential for life were born in the core of stars.[19] Our sun and solar system formed 4.6 billion years ago—long after the first stars formed more than 13 billion years ago. Perhaps a vast number of planets had already formed before Earth came into existence; many of these planets may have born life.[20] Life on Earth however formed between 3.5 to 3.8 billion years ago when a single-celled organism capable of reproducing itself first made its appearance. Nourished by geothermal energy, these first organisms were simple beings: single-celled entities called prokaryotes.

Life *may* be unique to Earth; at this point, we simply do not know.[21] A consensus opinion among astrobiologists is that life *as we know it* appears to be a planetary phenomenon, understood potentially to include planetary satellites, minor planets, and comets.[22] Extraterrestrial life, therefore, remains well within the realm of the possible, although thus far no evidence for its existence has been documented. As we will see later, this tremendously impacts both our understanding of life in general and our (humanity's) relationship to other living beings. Relatedly, the evolution of a living cell from the microscopic constellation of molecules on Earth may have happened more than once. If it did, scientists have yet to find any fossil evidence. What is known is that all life *as we know it* evolved from a common cellular ancestor. In fact, as Nurse writes, the deep commonalities in life's chemical foundations point to a remarkable conclusion: All living organisms on Earth are related at the molecular level.[23] From the smallest bacteria to the cells of the most complex species, all life on Earth is composed primarily of a few, key elements, notably carbon, hydrogen, oxygen, and nitrogen; roughly 98 percent of every living thing on Earth is made up by weight of these four main elements.[24] Whether living beings stand or swim or slither or fly, they are all related and evolved from a common ancestor that traces its lineage to nonliving matter.

Fossil evidence indicates that all Earth-bound living organisms were born of a process known as *abiogenesis:* the emergence of self-replicating organisms from nonliving chemical precursors. In this sense, as Lupisella writes, "there is no fundamental difference between life and non-life."[25] In other words, what we understand as dichotomies—life and nonlife, living and nonliving, organic and inorganic—are merely artificial separations of a cosmic continuum on which everything in the universe is related. Surprisingly, the fictional "elemental" creatures who inhabit the superhero genre are more closely like us than we ever imagined.

Life as we know it began as single-celled microbial organisms and, for the first 2.5 billion years of Earth's existence, these single-celled ancestors of ours reproduced clonally: one bacterium would divide into two separate bacteria, each containing the same genetic makeup. However, sometime around 1.2 billion years ago, these simple prokaryotic organisms developed a mutation. Rather than just a copy, these organisms paired with other microbes to create offspring, thus combining the genetic information of both parents. Sexual reproduction came into existence and the evolutionary possibilities of life expanded infinitely.[26]

Mutations are crucial to evolutionary transformations.[27] Genetic mutations can result from a variety of causes, for example, errors in copying DNA or damage from radiation; these mutations, in turn, can be beneficial, neutral, or harmful, depending on the nature of the variation and on the environment.[28] Species evolve—transform—because of small changes, or mutations, in inheritable traits and compete through natural selection; over time, species with traits conferring special advantages in a given environment thrive and reproduce more than those with less advantageous configurations.[29] Simply stated, natural selection is "the intensely creative process that has produced us—and the extraordinary diversity of living forms that surrounds us—from the millions of different microbe species to the fearsome jaws of the stag beetle, the thirty-meter tentacles of the lion's mane jellyfish, the fluid-filled traps of the carnivorous pitcher plant and the opposable thumbs of the great apes, including ourselves."[30]

Genetic mutations figure prominently in the superhero genre.[31] Most often, genetic mutations in superhero tales are *somatic*, meaning that these alterations occur *after* conception, that is, when the character is a child or young adult. Peter Parker (Spiderman), for example, is bitten by a radioactive spider, Matt Murdock (Daredevil) is blinded by a radioactive spillage, and Bruce Banner (The Hulk) is exposed to gamma rays. Similarly, the Fantastic Four are genetically altered after being exposed to cosmic rays. Less frequently, superheroes (and supervillains) are the result of mutations that occur "naturally" during conception. Both the X-Men and Inhumans are humans born with genetic mutations that manifest sometime after birth.

Throughout *Homo sapiens'* brief existence on Earth, alterations to the human body have occurred randomly, the result of genetic mutations and natural selection. Now, though, natural selection is rapidly being supplemented—some might say *supplanted*—with "artificial" selection, as advances in synthetic biology, bio-fabrications, nanotechnologies, artificial intelligence, affective computing, and genetic engineering are heralding a new dawn of *purposeful* evolution. Not unlike the artificial engineering of Steve Rogers (Captain America), biotechnological procedures are continually being introduced to "improve" on evolution and to remake the human into something more-than-human. As such, the materiality of the human has been transformed and, in the process, altered further the question of being human. Indeed, knowledge of life itself—from the scale of the molecule to that of the cosmos—has irrevocably altered our understanding of "the human being" and its place on Earth and in the universe. On this point, Rosi Braidotti observes, "In contemporary society the human has become a question mark. Who or what counts as human today?"[32] George Lakoff is no less insistent, asserting that "there is a revolution going on, a revolution in our understanding of what it is to be a human being. . . . As a society, we have to rethink what it fundamentally means to be human."[33] Consequently, following Elana Gomel, "If the concept of humanity becomes blurred, the foundation of ethics can and should be questioned."[34]

BEING HUMAN

You are looking at one hundred percent red-blooded Earth man.
—NICK FURY (*CAPTAIN MARVEL* [2019])

There was nothing predetermined about humanity's appearance. We are not the pinnacle or apex of evolution. The diversity and dispersal of humanity today is the result of millions of years of biological evolution. Accordingly, as Stephen Gould explains, *Homo sapiens* did not appear on the Earth because evolutionary theory *predicts* such an outcome based on themes of progress and increasing neural complexity; rather, humans arose as a fortuitous (for us) and contingent outcome of thousands of linked events, any one of which could have occurred differently and sent history on an alternative pathway that would not have led to consciousness.[35] Indeed, Gould counsels that "if a small lineage of primates had not evolved upright posture on the drying African savannas just two to four million years ago, then our ancestry might have ended in a line of apes that, like the chimpanzee and gorilla today, would have become ecologically marginal and probably doomed to extinction despite their remarkable behavioral complexity."[36] And yet, Gould concludes, "Our conventional desire to view history as progressive, and to see humans as predictably dominant, has grossly distorted our interpretation of life's pathway."[37]

Humans, in fact, are relative newcomers when it comes to life on Earth. Archaeological evidence suggests that *Homo sapiens* first walked the Earth approximately 200,000 to 300,000 years ago.[38] In other words, humans have been absent for more than 99.9 percent of the time that life has existed on Earth.[39] A curious thing happened on the road to becoming human, however. In many cultures and societies, humans began to see themselves separate both from other species and from Earth itself. The foundation of this belief—anthropocentrism—is built on the edifices of both religion and philosophy, notably within Western society, but not exclusively so. How—and why—this belief assumed dominance over the centuries has a tremendous bearing on humanity's ability to navigate an uncertain posthuman future.

So, what exactly does it *mean* to be human? This question has confronted theologians and philosophers for millennia and, no doubt, will continue to spark debate for centuries to come.[40] In Western society, religion has long held purview over the subject of being human. In the Bible, for example, "what is the human being?" is asked in both Job (7:17) and the Psalms (8:4).[41] In these and other parts of the Bible, human beings are part of God's creation and *man's* purpose in life is to discern and fulfill God's will. A dominant interpretation of Christianity holds that God made "man" in His own image; as such, "man" is morally separate from and superior to all other Earthly creations. The Bible is clear on this point: God created light and dark, the heavenly bodies, Earth and the planets, and all of Earth's plants and animals. Then, God created in His image Adam and, subsequently, Eve as Adam's companion. Adam was directed by God to name all of Earth's creatures, thus establishing mankind's dominance over them. Tellingly, in Genesis 1:28, Adam is directed to "Be fruitful and multiply, and replenish the Earth, and subdue it; have dominion over the fish of the Sea, and over the fowl of the air, and over every living thing that moves upon the Earth." As Lynn White explains, "God planned all of this explicitly for man's benefit and rule: No item in the physical creation had any purpose save to serve man's purposes."[42] Moreover, humanity was understood as occupying a distinct ladder on a "Great Chain of Being," a cosmological hierarchy with humans residing between the spiritual heavens and the bestial Earth. Humans were positioned to exercise dominion over all of God's other-than-human creations.[43] This mandate was no small order, for God commands human beings to "be masters of . . . all the wild beasts" (Genesis 1:26). Based on this Biblical passage, Paul Robbins and colleagues explain, "If people have indeed been commanded by an all-powerful deity to do this, then quite clearly, not only is subduing and dominating all of Earth permitted, it is, moreover, an ethical imperative. It is the right thing to do."[44] Effectively, nonhuman animals have no rights and all of Earth's bounty—land, water, plants, and nonhuman animals—exist to serve humanity. It follows, therefore, that if an entity—animate or inanimate, organic or inorganic—is deemed superfluous or harmful to humanity, it is regarded as expendable, subject to extermination.[45]

That said, other theologians and philosophers forward a competing thesis that Christianity offers a morality based on stewardship. Thus, while God has entrusted the Earth to humanity, this dominion is far from absolute: God remains the ultimate authority over his creations and he imposes limitations on what humans can rightfully do to nature.[46] Seen in this light, humanity has a moral obligation to serve as responsible stewards, as guardians of the Earth and all it contains. Regardless, both Biblical interpretations—dominion or stewardship—demonstrate an anthropocentric ethic that maintains a separation between humanity and the rest of the world.[47] In either case, humans generally treat nonhuman animals and the Earth as mere resources or commodities, lacking in all moral value.[48]

Other religions, including many Native American belief systems, present very different conceptions of humanity's relationship to Earth and its myriad other-than-human beings. In fact, several Indigenous cultures locate human beings in a larger social, as well as physical, environment. In these traditions, people belong not only to a human community, but to an enlarged community premised on mutual respect and obligations.[49] Historically, though, Western societies marginalized Indigenous knowledge and remained firmly wedded to the notion of mankind as the measure of all things; in turn, Western humanism exiled other-than-human beings "beyond the moral pale to become simply a non-human or sub-human counterpoint against which our concepts of humanity and moral standing could be erected."[50]

Long enshrined in the canons of Western political philosophy, the human being was heralded as the embodiment of rational thought, self-reflection, and self-interest.[51] Indeed, since the Enlightenment, Affrica Taylor writes, "human capacity to reason and exercise intentional agency has been celebrated above all else." She explains that the valorization of human rationality has provided the epistemological basis for separating our species off from the rest of the natural world and has affirmed the need for us to exercise our exceptional intelligence and agency in order to "improve" on or "fix" nature.[52] Posthumanism rejects the view of the human as exceptional, as a being separate from, and usually dominant over, other life forms.[53]

Still, it is necessary to ask, following Zakiyyah Jackson, "What and crucially *whose* conception of humanity are we moving beyond?"[54] This is important, for the "human" in Western thought has never been universal. Indeed, as scholars working in myriad fields, including but not limited to feminist studies, postcolonial studies, Indigenous studies, Black studies, and Queer studies, the Eurocentric and humanist idea of "the human" has been critiqued for its sexist, misogynist, and white supremacist assumptions.[55] As Braidotti explains, the human subject of Western political philosophy has always been a selective and exclusionary category that polices access to rights and entitlements.[56] The figure of the human, for example, was most often always and already conceived as a white, able-bodied, heterosexual man.[57] In turn, moral standing was reserved *only* for certain humans—those who conformed to this normative standard. Women and people classified as nonwhite, for example, were relegated to a second-class humanity, while nonhuman life-forms were excluded entirely and thus subjected to prejudice, discrimination, oppression, and harm.[58] Slavery is a clear expression of the denial of moral standing to human beings, as are misogynist and homophobic beliefs, attitudes, and practices. The consequences can be extremely injurious and even deadly. Some police departments have referred to certain people—including African Americans, Indigenous peoples, and sex workers—as nonhumans. When people of color were found murdered, for example, the reporting officers would describe the crime scene as having "No Humans Involved" or simply "NHI."[59]

"Ethical and political struggles," Gomel reminds us, "often attempt to expand rather than to breach the boundaries of humanity." For example, most political battles, Gomel expounds, "are waged around extending human rights to previously disenfranchised groups, such as women, gays, minorities, the disabled, and, later, to nonhuman animals and the environment itself. In our posthuman future, however, the question of extending human rights to more-than-human and other-than-human beings remains unanswered.[60]

In science fiction, mutants, androids, aliens, and myriad other "nonhuman" beings commonly function as surrogates for marginalized peoples and groups who are often dehumanized and subject to hatred, prejudice, discrimination, and violence. Accordingly, the popularity of the superhero genre in general, and of the MCU in particular, can be seen as part-and-parcel of our present posthuman condition. Indeed, our present condition—personified by a panoply of more-than-human and other-than-human beings—constitutes a moment in which "humans are no longer the most important things in the universe."[61] Left uncertain is humanity's ethical and moral response to a posthuman future.

TOWARD A BIOCENTRIC ETHICS

I am on the side of Life.
—VISION *(AVENGERS: AGE OF ULTRON* (2015])

Should microbes have rights? Should we follow Vision's lead and extend moral standing to all life, including those beings lacking sentience? The question is not far-fetched. As Charles Cockell explains, the idea that microbes might have some moral claims on us beyond their practical uses or instrumental value is not a new one.[62] In the 1970s, for example, Bernard Dixon questioned the ethics of eradicating the smallpox virus.[63] Dixon concedes, on one hand, the human toll exacted by smallpox but, on the other hand, questions the right of one species (*Homo sapiens*) to exterminate another species. Does this not constitute a form of genocide?

Notably, the moral dilemma posed by microbial rights centers on the observation that microbes have biological interests and, as such, whether this knowledge compels "us" to extend any consideration beyond "our" perceived practical uses of that organism.[64] During the Enlightenment, for example, it was widely assumed that only (certain) beings were deserving of moral standing.[65] This was the viewpoint expressed by René Descartes, for example, who concluded that nonhuman animals were insensible and irrational machines and existed almost exclusively for the benefit of science and (human) civilization.[66] Gradually, however, a minority of intellectuals challenged these anthropocentric attitudes. Moral standing was slowly but progressively extended to previously disenfranchised human beings, and then toward nonhuman animals and to the environment itself.[67] The English botanist John Ray, for example, argued that the belief that the natural world existed only for humanity's benefit was an unsupported conceit. Later, the writings of Charles Darwin especially eroded the long-held assumptions of human exceptionalism and anthropocentrism. As Nash explains, the more humans learned about nature, the more difficult it became to entertain the notion that the universe existed *for* humans. Progressively, biologists and other scientists forwarded the argument that humanity was not the master of nature so much as one of the members of the natural community.[68] Left unclear, however, was (and still is) how inclusive our moral community should be.

One commonly held position is that only *sentient* creatures should be extended moral standing. Here, a sentient being is defined by Matthieu Ricard as "a living organism capable of distinguishing between well-being and pain and between different ways of being treated—that is, between different conditions that are either favorable to its survival or harmful."[69] Since the 1970s, this position, that life-forms with the capacity of experiencing pain and pleasure have moral standing, has become the central tenet in much of the animal rights movement.[70] Peter Singer is notable in this regard. Singer argues that "if a being suffers, there can be no moral justification for refusing to take that suffering into consideration."[71] To do otherwise is be guilty of speciesism, defined by Singer as "a prejudice or bias in favor of the interests of members of one's own species and against those of members of other species, on the basis of species alone."[72] Thus, while Singer argues that all sentient beings are entitled to equal moral consideration, he does not advocate that all sentient beings receive equal rights. Rather, Singer adopts a principle of equal consideration, that is, the interests of all sentient beings should be considered and extended the same consideration as any other sentient beings. In other words, Singer argues that human beings should adopt a principle of equal consideration of interests whereby the interest of other *sentient* beings be considered as moral subjects and given equal weight when making ethical choices. This proposition, in practice, has been directed predominantly toward the industrial slaughter of farm animals and nonhuman animal experimentation, but also toward the captivity of nonhuman animals in zoos, aquariums, and related theme parks.[73]

The fulcrum on which rights are or are not applied often pivots on having *interests*. Joel Feinberg, for example, claims that "without awareness, expectation, belief, desire, aim, and purpose, a being can have no interest; without interests, he cannot be benefitted; without the capacity to be a beneficiary, he can have no rights."[74] Singer makes a similar argument, writing that "the capacity for suffering and enjoying things is a prerequisite for having interests at all, a condition that must be satisfied before we can speak of interests in any meaningful way." For Singer, it is nonsensical to speak of the interests of a rock, for some*thing* inorganic cannot suffer. Conversely, a mouse "does have an interest in not being tormented."[75] Does this suggest, therefore, that "we" treat mice and human beings equally, given that both species-beings are sentient and thus subject to equal consideration? Singer argues that we should not. Indeed, he concedes that the lives of some sentient beings have greater inherent value than those of others, and that the lives of humans are generally of greater worth than those of nonhuman beings.[76]

This position troubles many ethicists who challenge the innate sentientism of Singer's argument. In response, biocentrism has emerged as a counterpoint to sentientism.[77] Briefly stated, biocentrism holds that *all* living beings have *substantive* moral standing; it hinges on the presumption that all *living* organisms demonstrate their own unique ways of being.[78] As Ricard Matthieu explains, every species-being possesses and expresses "intelligence" in its own way and embodies particular abilities required to survive. Bats, for example, are guided by an extremely sophisticated sonar system and many migrating birds traverse the planet by orienting themselves by the stars. These are all feats of which humans—barring technological innovations—are incapable of performing.[79] Simply put, many nonhuman beings *understand* the world in which they live in their own, unique, nonhuman ways. Nonhuman beings are not simply *in the world*; they also are aware of *being in the world*. As such, we—as humans—should recognize and respect the simple fact of being in the world of all living organisms, from the smallest bacterium to the largest blue whale.

In opposition to more hegemonic beliefs centered on human exceptionalism and speciesism, biocentrism holds that "since any living organism is 'interested' in its well-being and continued existence, *prima facie* duties to respect an entity's well-being and continued existence are empirically tantamount to *prima facie* duties to respect an entity's 'interest' in well-being and continued existence."[80] As James Bridle explains, "From bonobos shaping complex tools, jackdaws training us to forage for them, bees debating the direction of their swarms, or trees that talk to and nourish one another. . . . the non-human world seems suddenly alive with intelligence and agency."[81] It follows, therefore, that "any and all living beings—humans, animals, plants—have moral standing."[82] However, so defined, an ethos of biocentrism necessarily privileges *living* beings and begs the question of what precisely constitutes life itself. Indeed, if—as the theory of abiogenesis argues—life emanated from nonlife, what do we make of robots and other forms of disembodied artificial intelligence? Could we not conceive these "objects" as intermediary, not-yet-living beings, as nascent "subjects" in transition and transformation from simply being to being alive?

As a counterpoint to anthropocentrism, therefore, biocentrism holds that not only do all life-forms have inherent value, but also that all living things have *substantive* moral standing.[83] This has tremendous purchase when interpreting the superhero genre, for a recurrent theme pivots on the social relations (and, by extension, the extension of moral obligations) between human beings and nonhuman beings. However, given that many biocentric positions center on the proposition that all and only living things have equal moral standing and deserve equal moral respect and concern, it is important to articulate more precisely the competing definitions and understandings both of life and death.[84]

On this point, life from a secular perspective may be understood in two different but not necessarily complimentary forms.[85] The first is biological or physiological; here, life is frequently (but not always) characterized by the beating of the heart, ventilation of the lungs, digestion, and nervous activity. As Steven Luper writes, a living organism is an entity that has a substantial capacity to maintain itself through particular vital processes, including homeostasis, metabolism, and reproduction; this, in turn, forms a basic understanding of life.[86] More precisely, as James Bernat explains, "an organism is a complex life form composed of individually living subunits, including cells, tissues, and organs. Each subunit is organized in a functional group, and is not merely a random aggregation of components. Thus, cells form functional subunits of tissues that in turn form functional subunits of organs. The interrelationships of the numerous hierarchies of functional subunits within an organism create an integrated, coordinated, functioning, and unified whole. This whole is the organism itself: the highest and most complex unit of life that subsumes all its living subsystems."[87] A second level refers to a sense of personhood, that is consciousness or sentience. Reasonably, these two different understandings of life may be paired with two understandings of death. Death, for example, can relate to the cessation of all (or some) of the biological functions that define life. Conversely, death may relate to "ceasing-to-exist" in any meaningful sense or, more properly, the loss of personhood.

The ontology of life and death, however, is considerably more challenging, particularly in our present posthuman era. With advances in science and technology, the ontological certainty of life and death has been fundamentally altered. Posthuman technologies, notably prosthetics, nanotechnologies, and genetic engineering, have transformed the basic premise of a "biological" paradigm of death, with corresponding implications for legal definitions of life and death. The designation of personhood, for example, is generally given to those entities who have moral and/or legal status; however, the notion of personhood is far from simple, in that—certainly in the Western context—not all humans are accorded personhood and not all persons are human.[88] As Jamie Morgan explains, "As a species, humans are creatures who are necessarily self-conscious, active, planning, creative, and cooperative (sometimes competitive but in ways that can require mutuality, dependencies, reliance, and perhaps trust). They are also able to change their own lived circumstances and the broader material environment precisely because of the nature of their being, which includes sociality."[89] Posthuman bodies challenge the presumed singularity (and superiority) of the human species and, by extension, the moral/legal standing of personhood; indeed, it is this challenge that mediates many struggles within the superhero genre. Notably, this provides an entry to consider the extension of moral obligations to "artificial agents," for example androids and robots, that is, to extend our moral obligations beyond biocentric framings to consider instead a "machine ethics" or "robot ethics."[90] Consequently, to expand biocentrist viewpoints in an increasingly posthuman universe, I turn to the ethics of Albert Schweitzer.

REVERENCE FOR (POSTHUMAN) LIFE

When does killing become the only way? We're S.H.I.E.L.D. Aren't we supposed to be better than that?

—ALPHONSO "MACK" MACKENZIE *(AGENTS OF S.H.I.E.L.D.,*
SEASON 5, EPISODE 20, "THE ONE WHO WILL SAVE US")

In the MCU, superheroes confront on a near-daily basis moral choices with life-and-death consequences. By-and-large, what makes a superhero heroic is that they have great powers and are thus *burdened* with great responsibilities. Certainly, they may make bad decisions but, in principle, they do

Matt Murdock (Charlie Cox) as Daredevil meets the Punisher, Frank Castle (Jon Bernthal) in *Daredevil* (2015).

not knowingly and unnecessarily inflict pain and suffering. When innocent people are harmed—and, frequently, even when villains are harmed, many superheroes—and those around them—express guilt and regret in the use of violent force. Arguably, both Marvel's *Daredevil* and *The Punisher*, both as stand-alone series and in crossover events, explore these tensions to a greater extent than do the films.

In Marvel's fictional universe, the contrast between Matt Murdock (Daredevil) and Frank Castle (the Punisher) reveals in stark clarity the difficulties of these choices. Murdock abhors violence, although he routinely is compelled to engage in brute force. He draws the line—some might say arbitrarily—when it comes to his refusal to take the life of another. In the second season of *Daredevil*, Murdock and Castle debate the morality of violence—notably after a titanic and violent clash between the two. In defending his position, Murdock declares, "I don't kill anyone." A stone-faced Castle is unmoved. "Is that why you think you're better than me?" he asks Daredevil. Murdock replies, "It doesn't matter what I think or what I am. People don't have to die." He continues, "I believe it's not my call, and it ain't yours either."[91] Murdock is a defense lawyer and believes in the rule of law. Certainly, his faith in the legal system waivers as the series progresses; however, his belief in the sanctity of life remains *mostly* unchallenged. "Redemption," he tells the Punisher, "It's real. And it's possible. The people you murder deserve another chance."[92]

Castle sees things differently. A former US combat soldier, Castle was trained—and authorized—to kill. However, Castle witnessed the brutal murder of his wife, son, and daughter and lost any faith in the justice system. Castle denies that he is "some crazy asshole" killing whomever he wants, but instead maintains that he only exacts retribution on people who deserve to die.[93] His first targets were the murderers of his family, but he soon redirects his violence toward what he perceives as a degraded society. In a heated exchange with Daredevil on the morals of taking life, Castle snarls, "This city, it stinks. It's a sewer. It stinks and it smells like shit and I can't get the stink out of my nose. I think that this world. . . . it needs men that are willing to make the hard call."[94]

The right to life sits at the core of moral dignity and is central to the narrative arc of the MCU.[95] As Jeff McMahan explains: "To kill a person, in contravention of that person's own will, is an egregious failure of respect for the person and his worth. It is to annihilate that which is irreplaceable, to show contempt for that which demands reverence, to assert a spurious authority over one who alone has proper authority over his own life, and to assume a superior position vis-à-vis one who is in reality one's moral equal. . . . Indeed, because killing inflicts the ultimate loss—the obliteration of the person himself—and is both irreversible and in-compensable, it is no exaggeration to say that it constitutes the ultimate violation of the requirement of respect."[96] Murdock refuses to kill; Castle accepts the burden and makes, from his point of view, the hard call.

Both individually and collectively, Murdock and Castle embody the moral tensions that permeate the MCU on the ethics of killing *humans*. However, Marvel's fictional world also presents important vistas from which to extend moral standing—and the right to live or die—into realms not currently possible in the real world. Notably, what of other-than-human beings? In the opening scene of *Guardians of the Galaxy Vol. 2*, Peter Quill, Gamora, Rocket Raccoon, and Drax fight and kill a monstrous, tentacled creature known as an abilisk. In *Guardians of the Galaxy Vol. 3*, however, Mantis befriends two abilisks, having discerned that the creatures are nonviolent and only attack if threatened.

If it is morally wrong to kill human beings, on what grounds is it morally acceptable to kill other species-beings, such as abilisks, but also inanimate or inorganic other-than-human beings? Singer, for example, refuses to extend moral standing to stones, on the presumption that these have no interests. And yet, our rapidly diversifying posthuman world is populated with robots, androids, and other forms of artificial intelligence. Should these beings receive equal moral consideration? These are crucial questions, all of which figure prominently throughout the MCU diegesis. To help us confront these questions, I forward the normative ethics of Albert Schweitzer.[97]

Born in the late nineteenth-century, Schweitzer was a polymath, a student of, among other things, theology, philosophy, and music. Above all, Schweitzer was a humanitarian and received in 1952 the Nobel Peace Prize for his work on the reverence for life. His ethos is presented concisely: "It is good to maintain and to encourage life; it is bad to destroy life or to obstruct it."[98] And to that end, Schweitzer writes of a "universal will-to-live," that is, that all living organisms, fundamentally, want to live.[99] As such, reverence for life, for Schweitzer, becomes the all-pervading principle that makes a true ethos possible and the foundation for all sound moral thought and action.[100]

Throughout his writings, Schweitzer repeatedly draws attention to the inadequacy of anthropocentrism and human exceptionalism as a normative ethos and seeks to widen the scope of ethics to incorporate concern for all manifestations of being.[101] Remarkably, Schweitzer does not limit the will-to-live to humans nor even to life as we know it—an extension of moral standing that will have tremendous purchase in our reading of the MCU. For Schweitzer, the will-to-live is discernable in "the flowering tree, in strange forms of medusa, in the blade of grass, in the crystal."[102] For Schweitzer, it is essential to live in a way that is fully attuned to the will-to-live of all living beings and, indeed, to contemplate also the preciousness and precarity of nonliving entities. In other words, our care and compassion is neither reducible nor restricted to life as we know it, but encompasses the totality of all that surrounds us. Accordingly, when Schweitzer affirms a reverence for life, he affirms the solidarity of all living things and the moral obligation of human beings who live in the midst of living things.[103] In this way, Schweitzer's reverence for life widens the moral circle to include more-than-human and other-than-human beings. This requires an awareness of *alterity*, which in this context implies an awareness of the existence and experiences of others—including nonhuman beings.

Schweitzer's reverence for life is a demanding ethos, one that he concedes can never be fully realized.[104] "The world is a ghastly drama," Schweitzer writes, "of will-to-life divided against itself."[105] In other words, the world is always and already rife with violence: Elk eat juniper plants; wolves eat elk. Humans cannot escape this drama. Humans require sustenance in order to survive and this requires the harm and suffering of other beings, including plants. "In a thousand ways," Schweitzer yields, "my existence stands in conflict with that of others. The necessity to destroy and to injure life is imposed upon me. If I walk along an unfrequented path, my foot

Clint Barton (Jeremey Renner) questions the ethics of violence in *Hawkeye* (2021).

brings destruction and pain upon the tiny creatures which populate it. In order to preserve my own existence, I must defend myself against the existence which injures it. I become a persecutor of the little mouse which inhabits my house, a murderer of the insect which wants to have its nest there, a mass-murderer of the bacteria which may endanger my life. I get my food by destroying plants and animals."[106]

How then it is possible to maintain such an ethos, as Schweitzer proposes, "in face of the horrible necessity" of our own will-to-live divided against itself?[107] For Schweitzer, our participation is unavoidable but it should never be unreflective. Throughout our everyday interactions with other species-beings, we are pressed by practical problems that demand judgement and action. These problems might originate in our need to eat or to protect ourselves in self-defense. As such, Schweitzer underscores the challenge of choosing between the ethical and the necessary. If the latter is chosen, for example, the necessity to kill for whatever reason, Schweizer writes, one must bear the responsibility and the guilt of having injured or taken the life of another being.[108] In the MCU, both Murdock and Castle reflect on the consequences of their respective decisions to take or spare the life of another human being. In the end, we may favor one position over the other—that is, to side with Murdock or Castle; regardless, their moral struggles provoke us to contemplate our own attitudes toward a reverence for life. Arguably, though, it is Clint Barton (Hawkeye) who best captures the essence of Schweitzer's ethics. When asked by Kate Bishop what was the best shot he ever made, Barton responds, "The one I didn't take."[109] In other words, Barton assumed the grave responsibility that arises when confronting another beings' will-to-live. Ultimately, as Schweitzer writes, everyone must decide for themselves in every instance whether to submit to the necessity of committing harm or of promoting life. Not surprisingly, Bishop will later confront the same ethical choice when she confronts the Black Widow, Yelena Belova.

For many critics, Schweitzer's ethos of a reverence for life is readily and easily dismissed, his meditations on life reducible to little more than a bio-utopian fantasy. And yet, to set aside a reverence for life is to miss the deeper meaning raised by Schweitzer—a meaning that permeates the narrative arc of the Marvel Cinematic Universe. Through our day-to-day activities we continually face ethical choices that impinge on the lives of other beings. A reverence for life serves potentially as a moral compass, made possible "only by coming to hear more and more plainly the voice of the ethical, by becoming ruled more and more by the longing to preserve and promote life, and by becoming more and more obstinate in resistance to the necessity for destroying or injuring life."[110] On this point, Schweitzer reveals the core of his bio-centered ethos: "No one must shut his eyes and regard as non-existent the sufferings of which he spares himself the sight."[111] In other words, we cannot remain indifferent to the harm and suffering of others; a reverence for life demands an ethos of responsibility toward the welfare of others. This awareness, however, is predicated on the paradox that our will-to-live is always and already divided against itself. Indeed, it is this awareness of the finitude of life that guides so many of the actions undertaken by superheroes, including characters in the MCU. In *Avengers: Age of Ultron* (2015), for example, Vision reflects on the necessity to destroy the malevolent robot, Ultron. "I don't want to kill Ultron," Vision explains, "He's unique. And he's in pain." However, Vision concedes that Ultron's pain "will roll over the Earth—so he must be destroyed." In this moment, Vision embodies Schweitzer's ethos of a reverence for life. His determination is not a rash judgement born of rage and vengeance, but instead a contemplative choice that reflects the pain of Vision having to participate in the extinction of another being. There is no glory, no satisfaction, but only the guilt of his pending action.[112]

In the end, Schweitzer promotes a reverence for life as aspirational: an ethos to find meaning in ethical care toward life itself. "The most fundamental principle of morality," Schweizer explains, is "that good consists in maintaining, promoting, and enhancing life, and that destroying,

Vision (Paul Bettany) personifies a reverence for life in *Avengers: Age of Ultron* (2015).

injuring, and limiting life are evil."[113] Accordingly, one's meaningfulness in life is expressed through one's reverence for life, that is, our devotion and our empathy toward the preservation of life most fully. By-and-large, the superhero's mission, Coogan writes, "is prosocial and selfless, which means that his fight must not be intended to benefit or further his own agenda. The mission convention is essential to the superhero genre because someone who does not act selflessly to aid others in times of need is not heroic and therefore not a hero."[114] If a broad consensus exists on the constitution of superheroes, what of their human counterparts? Indeed, unlike the cottage industry that has emerged to debate ad nauseum what it means to be a superhero, scant attention has been directed toward the question of being human in a world filled with superheroes and other nonhuman beings. In other words, how do humans identify *as humans* when surrounded by more powerful beings? And, more to the point, how far removed are ordinary humans in Marvel's cinematic universe from Schweitzer's ideal?

CONCLUSIONS

The world has gotten even stranger than you already know.
—NICK FURY (*THE AVENGERS* [2012])

You have a gift. You have power. And with great power, there must also
come great responsibility.
—MAY PARKER (*SPIDER-MAN: NO WAY HOME* [2021])

The ethics of reverence for life throw upon us a responsibility so
unlimited as to be terrifying.
—ALBERT SCHWEITZER (*THE PHILOSOPHY OF CIVILIZATION*, 320)

How, Marvin Meyer asks, can we understand the challenges of moral goodness, of evil and ethics, in the world? How can we see ourselves in the context of other living beings and, in turn, assume our responsibilities and act on them in a world of often painful and perplexing ambiguities?[115] In the MCU, much attention has been directed toward Peter Parker, Steve Rogers, Tony Stark, and all the other superheroes and supervillains. But what of the so-called ordinary humans who stand alongside or, in some circumstances, in opposition to these more-than-human and other-than-human beings? Whatever posthuman future awaits humanity, it is fallacious to assume that humanity as a whole will experience these transformations either equally or equitably. The fictional world of the MCU provides a glimpse into humanity's uncertain future.

To consider the MCU from a posthuman ethos is to take seriously the existence of all types of beings, including those that blur the boundaries of human and nonhuman, plant and animal, natural and artificial, organic and inorganic, terrestrial and extraterrestrial. The deep tension between the universality of reverence for life and the necessity of harming some to help others recurs frequently in the writings of Schweitzer and, I maintain, in the various narrative arcs of the MCU.[116] Indeed, the MCU can be read as an extended ghastly drama marred by spectacular displays of violence, death, and destruction.

That said, the MCU provides also an opportunity to explore, following Judith Butler, what makes for a grievable life.[117] This is significant insofar as the possibility of premature death is pervasive throughout the MCU. Thus, rather than a glorification of violence, the MCU opens new vistas to contemplate the meaning of life and, by extension, one's relationship to the death of others, for the act of mourning reminds us that our sense of self is intimately tied to the loss of others. In this way, loss has a transformative effect, in that hope is

In *Guardians of the Galaxy, Volume 2* (2017), Mantis (Pom Klementieff) displays an empathy not widely shared by her counterparts, including Peter Quill (Chris Pratt).

often inspired by grief.[118] Timothy Beal, on this point, suggests that hope is palliative, that hope is "about prioritizing quality of life over quantity of days or years alive. It's about learning to live with necessary pain and suffering, and, at least as importantly, alleviating unnecessary suffering."[119] Hope is necessary, if only because "the sheer fact that we are constituted in and by our relations with other beings . . . does not ensure that we will notice or grieve their loss, just as our shared humanity cannot guarantee that we will recognize, much less grieve, the loss of another human being."[120] Perhaps, in the end, we should emulate more the fictional character Mantis, an Insectoid-Celestial hybrid empath who can feel the emotions of other living beings.

In the chapters that follow, I do not position humanity as fundamentally special or exceptional—although several fictional characters do in the MCU perceive of themselves as such—but instead posit humanity as a starting point for thinking about more-than-human and other-than-human beings.[121] Ultimately, we are confronted with answering the question of who *we* as humans are. Too often, critiques of the superhero genre focus on the embodiment of the superhero as an autonomous, sovereign being; and yet, such a myopic view risks losing sight of the ethical relations between human beings and the myriad other-than-human and more-than-human beings that live and die in Marvel's posthuman cinematic universe. Peter Parker understands and accepts his responsibility toward others; it remains uncertain how "ordinary" humans will respond.

Chapter Two

MARVEL'S MONSTROUS OTHER

What you have is an out-of-control freak show.
—GENERAL TALBOT *(AGENTS OF S.H.I.E.L.D.,*
SEASON 3, EPISODE 20, "EMANCIPATION")

Mutants have long been a staple of both science fiction and superhero genres.[1] Informed both by the Darwinian theory of biological evolution and scientific experiments with radiation in the 1920s, science fiction writers soon considered the possibility of mutated beings as both protagonists and antagonists. An early example is Olaf Stapledon's *Odd John: A Story Between Jest and Earnest* (1935).[2] In Stapledon's story, the titular character, John Wainwright, is described as *Homo superior*, that is, a new, more capable species emerging from humanity.[3] Wainwright is a complex figure and serves to highlight the varied reactions of nonenhanced humans. Perhaps not surprisingly, neither humans nor their more-than-human counterparts are able to coexist, although fear, mistrust and animosity is most pronounced among ordinary humans. Indeed, Stapledon introduces a theme that continues through to the fictional worlds of Marvel, chiefly that mutated humans—*Homo superior*—literally embody the fragilities and vulnerabilities of being human. Always in constant reminder of their own mortality, humans reject Wainwright and the other more-than-human beings. *Homo superior* are monstrous in the eyes of humanity and thus prophesize an uncertain, dangerous future.

Our inherited word *monster* is drawn from the Latin *monstrare* (to "show" or "reveal") and *monere* (to "warn" or "portend"). At the most fundamental level, therefore, monsters appear as a warning and it is precisely this cautionary function that makes monsters meaningful. We should not, however, read "meaningful" as being inherently positive or agreeable. For something to be meaningful, that object must be important and perhaps also expressive of a particular state of mind or quality of life. The appearance of "monsters" in dystopian fictions, as such, often portends an ominous future, a clarion call to readers that society is on the wrong trajectory. The *representation* of the monster in science fiction narratives, including the MCU, is not primarily "about" real monsters. In other words, the appearance of so-called monsters calls attention to artificial boundaries erected between and within species. To paraphrase Gregory Claeys, what is crucial is not the monster but the idea of the monster, the concoction of what inspires hatred.[4]

The monster therefore is not simply a meaningless beast whose function is to run amok, incite terror, and kill indiscriminately; rather, the monster is both *meaningful* and provides meaning.[5] Judith Halberstam echoes this premise, noting that the monster always becomes a primary focus of interpretation and its monstrosity seems available for any number of meanings.[6] In other words, the monster's appearance marks a disturbance in society, a harbinger that all is not as it seems. As such, the monster is more than an odious creature of the imagination; it is a kind of cultural category. Sometimes the monster is a display of God's wrath, a portent of the future, a symbol of moral virtue or vice, or an accident of nature.[7] Regardless, the monster offers a space where society can represent and address anxieties of its time. Indeed, as Claeys remarks, "Monsters inhabit the primordial *terra incognita* of the Earth."[8] Monsters, as such, embody societal fears and desires, hatreds, and yearnings.[9] Simply put, as Donna McCormack posits, "The monster is a question: What kind of world do you want to live in?"[10]

When we use the metaphor of the monster to represent human beings, we devalue human beings, that is, we quite literally "dehumanize" them. Consequently, once a human is dehumanized, any cognitive dissonance is removed when they are deprived of rights they would otherwise be entitled if they were regarded as fully human.[11] In an increasingly posthuman or other-than-human world, we must acknowledge that dehumanization continues, albeit in a radically transformed way. Indeed, as Claeys remarks, to call someone—or something—as monstrous is "the first expression of radical otherness."[12] In short, to declare someone *or something* monstrous is to render them abject.

Abjection, Julia Kristeva explains, is not merely a "lack of cleanliness or health," but instead that which "disturbs identity, system, order."[13] In other words, abjection is inherently monstrous. The figure of the abject transgresses normative orders—including that which constitutes being human. As such, abject lives, following Kristeva, are "based on exclusion."[14] This exclusion may manifest as physical sequestration, for example, when the abject figure is relegated to segregated spaces. The abject figure is reprimanded to "know their place" in society. As such, the exclusion of the abject is always symbolic, for the mere existence of the monstrous figure warns of the inherent fragility and instability of established norms and the dominant social order.

Now, well into the twenty-first century, humans are increasingly confronted with abject beings who are no longer or never were fully human, for example, cyborgs, robots, and androids. And in the fictional worlds of superheroes, this list expands exponentially, to include gods, demi-gods, extraterrestrials, and so on. In science fiction, including its superhero variant, encounters with more-than-human or other-than-human beings, such as mutants, foreground issues of individual empathy and community formation. This is to say, belonging—to a group, a tribe, a nation, or even a species—entails a process of inclusion and exclusion. As Amartya Sen explains, a "strong—and exclusive—sense of belonging to one group can in many cases carry with it the perception of distance and divergence."[15] In other words, humans often express a clear preference for those they consider more or less like themselves. This preference can manifest in myriad ways, for example, in feeling "closer" to members of one's perceived racial group. Mutants especially, as often depicted in fictional accounts, are *not quite* human enough to be included in the folds of humanity. Effectively, more-than-human and other-than-human beings, including mutants, are excluded from the community of *Homo sapiens*.

Far from a neutral or apolitical concept, community implies a sense of security and solidarity, a togetherness often infused with a sense of purpose. As Stuart Aitken writes, if we assume that "community," with its vague and generally nurturing meanings, is usually something that people desire, then it must be something people do not yet have—or fear losing—in the way that they want.[16] Consequently, the promotion of "community," while lauded as something positive and inclusive, necessarily entails its opposite, that is, the practice of social-spatial exclusion. As David Sibley writes, "Who is felt to belong and not to belong contributes in an important way to the shaping of social space."[17] Sen captures this contradiction, noting that a "sense of identity can make important contributions to the strength and warmth of our relations with others, such as neighbors, or members of the same community, or fellow citizens, or followers of the same religion. Our focus on particular identities can enrich our bonds and make us do many things for each other and can help to take us beyond our self-centered lives."[18] On the other hand, Sen concedes that "a sense of identity can firmly exclude many people even as it warmly embraces others. The well-integrated community in which residents instinctively do absolutely wonderful things for each other with great immediacy and solidarity can be the very same community in which bricks are thrown through the windows of immigrants who move into the region from elsewhere."[19]

In-group preferences, however defined, can and do have powerful impacts on the lives of humans, more-than-humans, and other-than-humans. Prejudice, discrimination, and violence often are the material and legal consequences of perceived differences within and between species. The problem, of course, is that these categorizations *normally* rely on a singular attributes, notably race, gender, and sexual orientation. The extension of both empathy and moral standing, as such, is conditioned by binary thinking: us/them, male/female, human/nonhuman. Moreover, no one person is reducible to any one attribute. We are simultaneously many things: we are gendered, raced, classed, and so forth. And yet, we are constantly forced to suspend our intersectional subjectivities.[20] How and why these enforcements come into existence is often the operative set of questions in science fiction narratives and of humanity's response, collectively, to the existence of more-than-human and other-than-human beings. More often than not, it is the monstrous Other who threatens the status quo of human exceptionalism and human species supremacy.

In *The Archaeology of Knowledge*, the French theorist Michel Foucault proclaimed, "Do not ask who I am and do not ask me to remain the same: leave it to our bureaucrats and our police to see that our papers are in order."[21] And both in fictional worlds and the real world, bureaucrats have been happy to oblige, working to define and defend who counts as human and who is to be excluded. Here, Dean Spade's work on administrative violence against members of the trans community is especially important. Spade suggests that the administrative realm "may be the place to look for how law structures and reproduces vulnerability" for marginalized populations, including those routinely considered as abject.[22] Thus, in thinking about how categories and classification schemes are constructed, enacted, and enforced, we are better positioned to understand how "administrative norms" dehumanize and make monstrous the Other. In addition, Spade writes, a focus on administrative violence "expands our analysis to examine systems that administer life chances through purportedly "neutral" criteria, understanding that those systems are often locations where racist, sexist, homophobic, ableist, xenophobic, and transphobic outcomes are produced."[23] Consequently, "Administrative systems that classify people actually invent and produce meaning for the categories they administer, and that those categories manage both the population and the distribution of security and vulnerability."[24] Effectively, to fully grasp how humanity will respond to more-than-human and other-than-human beings, we must redirect our focus away from individual actions of prejudice, bigotry, discrimination, and violence to consider how bureaucratic practices and material documents enable and legitimate these actions in a posthuman future. In short, we must confront more precisely how mutants are administratively made monstrous.

In Marvel's fictional world, mutants are not "natural" monsters but instead made monstrous through the anxiety experienced and antipathy expressed by so-called normal humans. Unlike the mutated monsters that populated comics and films in the 1950s—think *Creature of the Black Lagoon*—Marvel's mutants are ordinary humans, that is, until their genetic differences appear during adolescence. As Stan Lee explains in creating the X-Men—a superhero team of genetically mutated humans—"I wanted to spotlight a group of innocent people who were feared and shunned and later hunted and persecuted. I wanted to show how anyone, no matter how blameless, can be victimized if the fates so decree."[25] Indeed, Marvel's mutants are routinely and repeatedly marginalized by an intolerant society that fears anything that deviates from normality.[26] Perceived as "outcasts or "freaks" by ordinary humans, mutants characteristically face oppression, discrimination, and prejudice. Mutants therefore are made monstrous not by their own actions, but instead by those humans who fear difference. Monsters, Kim Newman reminds us, "are rarely as simple as they at first seem."[27]

To explore this uncertain future as portrayed in the MCU, in this chapter, I focus primarily on Inhumans, a species of genetically mutated humans who feature exclusively in the ABC television series *Agents of S.H.I.E.L.D.* This approach, I admit, is not without risks, foremost being that the series is not currently considered part of the MCU canon.[28] Arguably, though, it is *Agents of S.H.I.E.L.D.* that has (to date) confronted most openly the concept of monstrosity, in that the presence of Inhumans challenges the status quo of human exceptionalism and species-supremacy.[29] Indeed, it is my argument that Marvel's Inhumans, as monstrous beings, provide a cautionary reminder of humanity's fear of inhabiting and maintaining control over, a more-than-human future, a future that no longer guarantees the supremacy of being human.[30] As such, in the violent persecution of mutated human beings, humanity collectively demonstrates an inability to pursue an ethics inspired by a reverence for life, an ethics centered on compassion and care toward other-than-human beings who harbor nothing more than a simple will-to-live. For following Albert Schweitzer, a person is truly ethical only when they obey the compulsion to help all life which they are able to assist, and to refrain from injuring anything that lives.[31]

32   CHAPTER TWO

MARVEL'S SUPER-POWERED BUREAUCRATS

Are you really that dense? S.H.I.E.L.D. monitors potential threats.
—NATASHA ROMANOFF TO BRUCE BANNER

Captain America is on threat watch?
—BRUCE BANNER

We ALL are!
—NATASHA ROMANOFF *(THE AVENGERS* [2012])

The term *bureaucracy* originated as a protest against incompetence, ineffi-
ciency, and excessive governmental regulations.[32] In the late seventeenth century, Jean-Baptiste Colbert was Comptroller General of Finance in France.
Subsequently, Colbert initiated a series of rules that were to be applied con-
sistently throughout the realm. Colbert believed that the uniform application
of rules would demonstrate fairness and equality of the government. Nearly
a century later, however, France's Administrator of Commerce Jean-Claude
Marie Vincent de Gournay derided the ever-increasing rules and regulations
as harming business activity. Thus, to "symbolize the idea that rule-makers
and rule-enforcers who did not understand or care about the consequences
of their actions were ruining the French government, he coined the sarcastic
term *bureaucratie*—government by desk."[33]

Documents are essential for bureaucratic rule. Indeed, recent scholarship
has shown how bureaucratic documents are produced, used, and experienced
through procedures, techniques, aesthetics, ideologies, cooperation, negotia-
tion, and contestation.[34] In short, this work has effectively detailed the importance of understanding "the materiality of information."[35] Two functions are
particularly notable. First, "documents promote control within organizations
and beyond not only through their links to the entities they document but
through the coordination of perspectives and activities."[36] This implies that
we take seriously the "objects" of politics, including "both the militarized
and the mundane."[37] Second, bureaucratic documents entail a "generative
capacity" in that they bring *things* into existence.[38] Marie-Andrée Jacobs,
for example, details how documents, but especially consent forms, generate
"form-made persons." Jacobs elaborates that while documents "answer the
bureaucratic needs for efficiency and for comparability of documents," they
also "make political subjects visible." In turn, these subjects may more readily
be "archived, classified, measured, compared, and controlled on a mass scale."[39]

In short, documents "are central to how bureaucratic objects are enacted in practice."[40] A focus on bureaucratic documents in the MCU, consequently, calls to question the materiality of power as it is administered behind the scenes, for here we see how seemingly mundane technologies—files, charts, lists, and records—become the means that enable nonenhanced humans to control super-powered, more-than-human, and other-than-human beings.

Bureaucratic documents are not neutral, but rather politically charged objects that exhibit a surprisingly high degree of agency, for it is through documents that subjects come into being and subsequently become subject to disciplinary control.[41] That said, administrative systems often *appear* "neutral," especially when discrimination has been framed as a problem of individuals with bad intentions who need to be prohibited from their bad acts by law.[42] This is, for example, the fundamental premise of most hate-crime laws that render discrimination and prejudice as expressions of personal animosity instead of systemic forms of hatred. And yet, as Spade explains, it is often the case that "the very administrative systems that determine what populations the law exists to promote and protect are the greatest sources of danger and violence."[43] In the United States, this is seen clearly in the violence directed toward trans people, persons of color, and religious minorities; in the fictional world of the MCU, it is seen in the violence directed toward those marked as more-than-human or other-than-human. Simply put, nonhuman beings are produced and reproduced by administrative fiat as "dangerous" subjects that require governmental containment and control.[44]

In the MCU, more-than-human and other-than-human beings are often portrayed by humans as being intrinsically dangerous. Indeed, the enhanced being's potential for danger is always and already offset against the vulnerability of "ordinary" humans. This means, in practice, that more-than-human and other-than-human beings are routinely subjected to prejudices, bigotry, and discrimination based mostly on fear. Superheroes, as such, are often excluded from humanity's moral community, not for what they have done but instead for what they might do. Charged "forever guilty" of harms not yet committed, the more-than-human or other-than-human being is subjected to bureaucratic rule, including continual surveillance and monitoring, detainment, and possible extermination.[45]

Governmental efforts to classify, monitor, and control "dangerous individuals" are frequently supported through the use of security discourses. Here, security discourses do not so much respond to objectively existing threats, but instead these discourses themselves constitute and actualize dangers, select risks, and prioritize threats.[46] In other words, following Marieke de Goede and Samuel Randalls, "threats do not exist prior to practices of articulation and identity" but rather "it is through modes of representation and imagination that threats are brought into being, and are perceived as endangering particular communities and as demanding particular forms of social action."[47] This practice is recurrent in the superhero genre and the MCU is no exception, for alongside the supervillain, the superhero embodies also perceived threats to the security of humanity's future existence. It matters little if enhanced individuals actually do commit harm; the threat (as perceived by humans) is that other-than-human and more-than-humans *are always and already* capable of committing harm. Moreover, harm need not be intentional. The collateral loss of life and of property to humans (and other nonhumans) are legitimate concerns whenever and wherever superheroes battle supervillains. It is hardly surprising, therefore, that humans attempt to defend humanity *against* Earth's mightiest defenders.

The connective administrative tissue of the MCU is the acronymically challenged agency known as S.H.I.E.L.D.[48] When first introduced in the comic book *Strange Tales #135* (August 1965), S.H.I.E.L.D. referred to the Supreme Headquarters, International Espionage, Law-Enforcement Division. Subsequently, the organization was recast as the Strategic Hazard Intervention Espionage Logistics Directorate before assuming its current identity as the Strategic Homeland Intervention, Enforcement and Logistics Division.[49] Over the years, S.H.I.E.L.D.'s terminological identity crisis has not gone unnoticed and is the source of many inside jokes. In an early episode of *Agents of S.H.I.E.L.D.*, for example, Deputy Director Maria Hill asks fellow agent Grant Ward what the organization stands for. Ward quips, "It means someone really wanted our initials to spell out 'shield.'"[50] Hill, of course, was not actually asking what the acronym stands for, but instead questioning the more pressing issue, of what the agency *stands* for in terms of ideology and practice.[51]

Not surprisingly, Nick Fury, the longtime director of S.H.I.E.L.D., provides the most cogent description of S.H.I.E.L.D.'s prime directive in just one word: "protection." Fury elaborates: "Sometimes, to protect one man against himself. Other times, to protect the planet against an alien invasion from another universe. It's a broad job description."[52] And indeed, throughout the MCU digenesis, government officials and agents rationalize and justify their actions, most often in the name of national (or planetary) security and protection. Regardless of personal intention, however, the enforcement of governmental edicts instantiates what Spade refers to as the "circulation of norms."[53] In this sense, a circulation of norms produces and reproduces particular ideas that certain populations must be "protected" from other, often *monstrous*, beings—that is, those outside the norm—who pose a threat to the security and well-being of "normal" human beings. In the posthuman world of the MCU, enhanced beings—including superheroes—are routinely portrayed as being outside the norms of humanity and thus embodying an existential threat to humanity writ large. The ongoing circulation of norms achieves the overall purpose of producing security for some populations and vulnerability for others.[54] Of paramount importance to the circulation of norms in the MCU is the use of lists or indices to keep tabs both on superheroes and supervillains.

Long-held as the rudimentary raw material of scholarly analyses, lists have now attracted critical attention as objects in their own right.[55] As Cornelia Vismann explains, "individual items are not put down in writing for the sake of memorizing spoken words, but in order to regulate goods, things, or people. Lists sort and engender circulation."[56] It is necessary therefore to account for the particular operations which the format of the list enables. List-making, whether derived through the assemblage of digitally acquired information or through biographies written with pen and ink, is not only a problem of selection, but also is a "transformative and performative practice that produces the items which the list will comprise."[57] Urs Stäheli maintains that we have to understand the "mediality" of a list, in other words, that we recognize that the list must come into existence before someone may be rightly or wrongly included on the list.[58] As a case in point, it only becomes possible to list a person as a threat if we have already determined the existence and presence of threats. In this way, lists produce their own reality and, by extension, the actuality of list-able men and women becomes a self-evident fact. Approaching lists in this way invites us to think differently about how diffuse security powers are created, expanded, and sustained.[59] To this end, and departing from an instrumentalist perspective that understands lists as simply the compilation or representation of preexisting information, lists are more robustly understood as "inscription devices" that produce specific material, political, and legal effects—including the circulation of norms.[60] As such, we need to consider in our posthuman future more precisely the type of work that lists perform as a specific technique of government.[61] Security lists, for example, may be viewed as knowledge practices and modes of legal ordering.[62] Consequently, following de Goede and Sullivan, "situating the materiality of the listing process at the center of analysis helps to bring the specific legal ordering capabilities of lists—that is, the ways they work to constitute law and establish new modes of legal transmission—into clearer view."[63]

In Marvel's comic book series, the US government continually introduces legislation (and corresponding institutional apparatuses) to monitor more-than-human and other-than-human beings, chiefly the Mutant Registration Act and the Super-Human Registration Act. These acts are thus far absent from the MCU. Instead, the cinematic franchise introduced the Sokovia Accords to perform a similar function. Composed of a set of legal documents, the Sokovia Accords constitute a legal instrument designed to register and regulate more-than-human and other-than-human beings. Broad in scope, the Accords apply to *all* enhanced beings, including those who work for government agencies such as S.H.I.E.L.D. The main targets, however, are the Avengers.

The catalyst for the Sokovia Accords is traced to a mission conducted in Lagos, Nigeria when several HYDRA operatives, including Brock Rumlow, attack the Institute for Infectious Disease in an effort to steal material to make biological weapons. The Avengers thwart Rumlow but, in the ensuing melee, twenty-six civilians are killed, including several Wakandan citizens engaged in charitable work. The mission is a public relations disaster and Wakanda's king, T'Chaka, spearheads a multinational effort to curb the exploits of the Avengers. In the United States, officials are mostly sympathetic to the Avengers, but concede that they are largely powerless to prevent the Accords from being implemented. Indeed, several politicians view the Accords as a necessary evil. US secretary of state Thaddeus Ross, for example, acknowledges that "the world owes the Avengers an unpayable debt." To the Avengers, he says, "You have fought for us, protected us, risked your lives" That said, Ross continues, "While a great many people see you as heroes, there are some who would prefer the word vigilantes."[64] In fact, Ross explains, many governments throughout the world simply see the Avengers as "a group of US-based, enhanced individuals who routinely ignore sovereign borders and inflict their will where they choose and who, frankly, seem unconcerned about what they leave behind."[65] The Sokovia Accords, signed by representatives of 117 countries, stipulate that the Avengers will no longer constitute a private organization and can operate only under the supervision of a United Nations panel. Any action not sanctioned by the UN would constitute a criminal act and the offending Avenger would be subject to criminal prosecution. In addition, the Accords lead to the creation of new apparatus, the Joint Counter-Terrorism Task Force and the Joint Counter Terrorist Center (JCTC), the latter headquartered in Berlin and manned by CIA agents Everett Ross and Sharon Carter.

The Accords, in legal terms, classify the Avengers as dangerous subjects; they are, effectively, adjudged forever guilty of future harms and thus require judicial and police oversight. Notably, the other-than-human Vision accepts this premise. "Our very strength incites challenge," he explains, reflecting also that "challenge incites conflict. And conflict. . . . breeds catastrophe."[66] By extension, the establishment of the JCTC marks a significant expansion of an unfolding multinational preventative state, a law-enforcement agency

US secretary of state Thaddeus Ross (William Hurt) attempts to curb the power of more-than-human beings in *Captain America: Civil War* (2016).

entrusted to watch the watchers. On a personal level, however, the Accords sow division within the Avengers, as team members debate and take sides regarding the merits of the legislation. Tony Stark, the former arms manufacturer, supports wholeheartedly the regulations. "We need to be put in check," Stark argues, concluding that "if we can't accept limitations, we're boundaryless, we're no better than the bad guys." Steve Rogers is skeptical, warning that the Accords undermine the autonomy of the Avengers. The United Nations, Rogers asserts, is "run by people with agendas and agendas change." As such, Rogers worries that the Avengers might become an instrument of governmental maleficence. "What if this panel sends us somewhere we don't think we should go," Rogers asks, "what if there's somewhere we need to go and they don't let us?" In the end, Rogers concludes, "We may not be perfect but the safest hands are still our own."[67]

It is interesting to consider the origins of the Avengers in light of the Sokovia Accords. As noted earlier, a central function of S.H.I.E.L.D. (which, by the time of *Captain America: Civil War*, was disbanded following its infiltration by HYDRA) is to operate behind the scenes in order to keep *humanity* safe from potential threats. Nick Fury, however, realized that many threats to humanity were embodied by a strange assortment of mutants, androids, and aliens, that is nonhuman beings that humans were ill-equipped to confront on their own. Consequently, Fury, together with Stark, concocted the "Avengers Initiative," a covert program envisioned to assemble a team of more-than-human and other-than-human beings to protect Earth—and especially its human inhabitants—from other, similarly enhanced beings. From the outset, there was no

Nick Fury (Samuel L. Jackson) is the mastermind of the Avengers Initiative, a covert program to protect Earth from more-than-human and other-than-human beings.

governmental oversight of the Avengers; it constituted an extrajudicial arm of a covert agency that also frequently operated outside the law. The Avengers, effectively, functioned as judge, jury, and executioner. The Sokovia Accords were drafted specifically to target the unlimited authority of the Avengers but, in doing so, led to unforeseen consequences, including the persecution of a genetically mutated species of human-beings: the Inhumans.

INHUMANS

> Hold on, are you saying that I'm an alien?
> —DAISY JOHNSON *(AGENTS OF S.H.I.E.L.D.,*
> SEASON 2, EPISODE 5, "A HEN IN THE WOLF HOUSE")

From the 1950s onwards, countless fictional worlds both in print and on-screen have been populated by genetically mutated humans and their non-human counterparts. It was only a matter of time before Marvel Comics followed suit. In 1963, Stan Lee and Jack Kirby introduced the X-Men to Marvel's line-up of superheroes and, after a faltering start, the team of mutated humans became one of the most popular franchises in the Marvel repertoire.[68] In a 2014 interview, Lee recalls how the concept of the X-Men came about. Building on the popularity of the Fantastic Four (first appearing in 1961), Marvel wanted to introduce another team of superheroes. The initial challenge was to identify their individual superpowers; the next and biggest challenge was to establish their origins. To date, most superheroes (and supervillains) obtained their powers by accident. The human members of the Fantastic Four, for example, were exposed to cosmic rays; Peter Parker (appearing in 1962) acquired his superpowers after being bitten by a radioactive spider. Bruce Banner (appearing in 1962) became The Hulk after his accidental exposure to gamma rays. Other superheroes obtained their enhanced powers by extensive training, such as Dr. Strange (appearing in 1963) who underwent extensive training in the "mystical arts." As for the X-Men, Lee explains, "I had to figure, how did they get their powers? And they were all separate people that weren't connected to each other, so I knew that would be a helluva job. And I took the cowardly way out, and I figured, hey, the easiest thing in the world: They were born that way. They were mutants. So I thought that would be one way to get around having to find new origins."[69] Whether Lee actually took the cowardly way out or not, the idea took root and the mutants charted a new path to reflect on the meaning of being human in an other-than-human world. In fact, from their inception, the central theme of the X-Men franchise has been the condemnation of prejudice and varied forms of social and political supremacy.[70] In a 1967 interview, for example, Lee affirms that "the more I realize that people are to some degree affected by what we write, the more I'm aware of the influence we have, the more I worry about what I write. ... I think the only message I have tried to get across is for Christsake don't be bigoted. Don't be intolerant."[71]

Through the introduction of mutants into their comic books, Marvel writers and artists could confront head-on themes of rejection and societal inequalities through the use of the "mutant metaphor."[72] Articulated most clearly by Joseph Darowski, the mutant metaphor "can potentially be applied to anyone who feels different from the dominant culture around them."[73] Certainly, as Miller elaborates, "the X-Men and mutants in general can be seen as metaphors for any number of minority or marginal groups."[74] And, in fact, Marvel has utilized mutants as the abject Other for a wide range of groups that have been persecuted, including ethnic minorities, Jewish and Islamic Americans, immigrants, and LGBTQ people.[75] In doing so, Ramzi Fawaz explains, Marvel has provided a foundation for reimagining the genetically mutated superhero as a way to dramatize the politics of inequality, exclusion, and difference.[76]

For contractual reasons, however, Marvel Studios was unable to incorporate the widely popular mutants—specifically, the X-Men—into the expanding MCU. In the 1990s, Marvel sold the rights to their mutant characters, including the X-Men, to 20th Century Fox; consequently, when the MCU began to take shape in the early 2000s, Marvel was unable to even reference mutants in the films. Screenwriters, producers, and directors, subsequently, were compelled to work with a lesser-known species-being, the Inhumans, alongside a menagerie of cyborgs, androids, and extraterrestrials, to fill the void.[77]

As comic book characters, the Inhumans first appeared in the comic book *Fantastic Four* #45 (December 1965), but remained tangential at best in the world of Marvel. That said, as the MCU took shape, studio executives were optimistic about the incorporation of Inhumans into their fictional cinematic world. Kevin Feige, President of Marvel Studios, for example, asserted, "We really do believe that the Inhumans can be a franchise and perhaps a series of franchises unto themselves. . . . They have dozens of different power bases, they have an amazing social structure." He explained that "It felt time to continue to further refine and further expand what the cinematic universe is all about."[78] In fact, so great was the media hype that Feige announced plans for a subsequent film centered on Inhumans.[79]

Building on MCU's premise that everything is connected, *Agents of S.H.I.E.L.D.* included many crossover references, including the on-screen appearances of several notable MCU characters such as Phil Coulson, Nick Fury, Maria Hill, Jasper Sitwell, and Lady Sith. Moreover, the show aligned with events transpiring on the big screen, with explicit references to the Battle of New York, the Sokovia Accords, Project Insight, and HYDRA's infiltration of S.H.I.E.L.D. And to further cement *Agents of S.H.I.E.L.D.* into the wider corpus of the MCU, in 2015 Marvel released *Inhuman* and *Uncanny Inhumans*, two comics produced explicitly to compliment *Agents of S.H.I.E.L.D.*[80] Clearly, as Lewis Glazebrook writes, "the show undoubtedly began as a canon inclusion to the MCU thanks to a variety of wider franchise connections." It was during this time, also, that Marvel Television expanded its programming, producing a number of television series with Netflix, including *Daredevil, Jessica Jones, Luke Cage, Iron Fist, The Punisher*, and the crossover miniseries, *The Defenders*.[81] As Feige explained, many of these developments were inspired by the success of *Agents of S.H.I.E.L.D.*[82] Indeed, in a 2024 interview Ming-Na Wen, the actress who plays S.H.I.E.L.D. agent Melinda May, commented that "We were the first. So you would think, like, we should get some accolade for helping to launch Marvel into [the television genre]."[83]

And yet, even as future crossover events were in preparation, the relationship between the Inhumans and the wider MCU began to unravel, a separation precipitated not by events in the fictional world, but instead from a bitter feud within Marvel Entertainment.[84] In 2015, Marvel Television and Marvel Studios separated and, in turn, the latter company would assume oversight and production of all live-action series that would appear in the MCU.[85] *Agents of S.H.I.E.L.D.* continued, driven in part by a loyal fan base, although, as the seasons progressed, any ties between the television series and the MCU became more tenuous.[86] Other related projects floundered. The once-ballyhooed film centered on the Inhumans was downscaled to a television series. It debuted in 2017 and was canceled after only eight episodes, derided by fans and critics alike. As Jordan Iacobucci writes, "Most viewers understandably pretended that *Inhumans* never happened, and they have Marvel Studios support on this front."[87] If the failure of the series undermined any possibility that the Inhumans would make an appearance in Marvel's widely popular cinematic universe, the company's acquisition of 20th Century Fox in 2019 almost certainly guaranteed their extinction, for the merger opened the door to introduce mutants into Marvel's fictional world. Consequently, Marvel has steadfastly worked to distance the unfolding MCU from anything remotely connected to Inhumans, going so far as to entirely rewrite the origin stories of characters who appear in the canonical MCU. In the comics, for example, Ms. Marvel is an Inhuman; in the MCU, however, she is refigured as a mutant.[88] Also, in 2018, Marvel Comics released a five-issue comic book story titled "Death of Inhumans."[89] In the series, thousands of Inhumans are slaughtered by the Kree, an extraterrestrial species. Adding insult to injury, a character from the canceled *Inhumans* series, Black Bolt, briefly reappeared in the MCU's *Doctor Strange in the Multiverse of Madness*, only to die a quick and gruesome death. As Catherine Mora notes, some fans speculated that Black Bolt was killed because of Feige's dislike of Inhumans.[90] To date, the Inhumans have been almost totally eclipsed from the MCU canon and the future of several popular characters, notably Daisy Johnson (Quake), remains in doubt.

For fans and scholars alike, the Inhumans pose a conundrum, in that the fictional characters and, in fact, their entire species, is one of inclusion and exclusion. And it is, in part, because of this lurid backstory I remain drawn to the Inhumans as depicted in *Agents of S.H.I.E.L.D.* Indeed, there is a curious parallel at play, as an entire species-being—marginalized and persecuted on-screen—is being erased from the so-called interconnected canon of the Marvel Cinematic Universe.

In the pilot episode of *Agents of S.H.I.E.L.D.* we find Phil Coulson assembling a team of S.H.I.E.L.D. agents, including deputy commander/pilot Melinda May, field agent Grant Ward, and scientists Leo Fitz and Jemma Simmons. Subsequently, Coulson enlists a computer hacker, Skye, to work

Daisy Johnson (Chloe Bennet), Grand Ward (Brett Dalton), Jemma Simmons (Elizabeth Henstridge), Leo Fitz (Iain De Caestecker), Phil Coulson (Clark Gregg), and Melinda May (Ming-Na Wen) assemble in *Agents of S.H.I.E.L.D.* (2013).

The Inhuman Daisy Johnson (Chloe Bennet) encounters bigotry and violence from "ordinary" humans.

with S.H.I.E.L.D. as a "consultant." Skye, though, is distrustful of the US government and its covert apparatuses, most especially S.H.I.E.L.D. Played by Chloe Bennett, Skye becomes the "audience's point-of-view character" and assumes prominence as the series progresses.[91] Indeed, Skye serves as "a stand-in for our own questions and ambivalence around the work of intelligence and security agencies" in that her fictional role "is precisely to mirror our own concerns and to act as an inroad into the workings of the security agency and its activities." Moreover, given that Skye is committed to uncovering and unmasking the truth about "the weirder world" where she now lives, Skye personifies the uncertainties of a posthuman universe.[92] Audiences, however, are thrown a loop in the second season when Skye—who now refers to herself by her given name, Daisy Johnson—undergoes a metamorphosis and is transformed into the Inhuman superhero, Quake.[93] Subsequently, it is through Daisy Johnson (Quake) we witness also the prejudices, bigotry, fear, hatred, and violence directed at Inhumans by "ordinary" humans, including—at times—members of S.H.I.E.L.D.

In the X-Men franchise, genetic mutations occur naturally: Some humans are born with the x-gene mutation while the majority of humans are not. Inhumans are different, in that their mutation is the result of extraterrestrial experiments conducted thousands of years ago. An alien species-being, the Kree, sent research teams to various planets for the purpose of creating a mutated species of enhanced soldiers to fight in a protracted inter-planetary war. As Vin-Tak, a Kree alien, explains, "We needed killers."[94] Most of these efforts failed, except on Earth, where some humans were able to survive the medical procedures. As such, when humans who carry Kree DNA are exposed to mist derived from Terrigen Crystals, their mutations are triggered during a process known as Terrigenesis. Exposure to Terrigen crystals, however, is fatal for humans who do not carry the mutated gene.

For Inhumans, the process of Terrigenesis marks a coming-of-age. As Raina, an Inhuman, reflects, "This [transformation] doesn't destroy. It gives life—new life. We finally get to find out what we become."[95] When exposed to Terrigen mist, the genetically mutated human is encased in a clay-like substance and emerges, chrysalis-like, with their enhanced powers activated, a transformation described by Daisy Johnson as a "biomorphic event."[96] The change from human to Inhuman is thus quite sudden. As Lincoln Campbell, another Inhuman, explains to Johnson, "Imagine a thousand years of evolution taking place instantaneously. That's what happened to you after the mist."[97]

Following Terrigenesis, some Inhumans appear much as they did before, that is, there is no outward physical change in their appearance. Johnson, for example, looks the same physically as she did prior to undergoing Terrigenesis. However, she soon discovers that she has the ability to manipulate the naturally occurring vibrations of objects, including Earth. Indeed, Johnson can literally cause the Earth to quake. Campbell, able to control electricity, appears also as an 'ordinary' human and is able to integrate in human society. Raina is different. After her metamorphosis Raina develops the power of clairvoyance. Unlike Johnson and Campbell, though, Raina's physical appearance is dramatically altered. Her skin is covered with razor-sharp thorns, barbs, and quills; her fingers extend as eagle-like talons; and her golden eyes are snake-like.

Inhumans struggle to understand their transformed selves. Of the process, Johnson says, "when I changed I felt like a monster. I needed help."[98] In time, and with the assistance of other Inhumans, such as Campbell, Johnson realizes that she is not "some horrible thing," but instead has "a purpose." Johnson's transformation, and that of all Inhumans, is thus reflective of real-world transformations that happen all the time. By illness or accident, people are transformed, physically, mentally, and psychologically. Society, however, can be intolerant and bigoted, both fearful and frightened of change.

For Johnson, becoming Inhuman did not mean that her life was over but instead was "just getting somewhere" and that perhaps there was a "place to belong" in human society.[99] Johnson later explains to S.H.I.E.L.D. agent Bobbi Morse, a human with no superpowers, "We're only part alien, and it's not a disease—it's an awakening." She continues, "I mean, for me, Terrigenesis made me who I was supposed to be. I was given a gift, and I use it to stop bad people from doing bad things."[100]

The struggle to accept oneself is made more difficult if the Inhuman no longer looks human. Raina, notably, attempts to commit suicide after her Inhuman self is revealed. "I wanted to be set free," Raina confides to Johnson, "but now I can't even bear the light of day. I dread being awake, but my sleep is filled with such horrible nightmares. Children are so afraid of monsters. They should know it's worse to be one."[101] Most humans do look upon Inhumans as something monstrous and it is this prejudice that compels Inhumans to separate themselves from human society—to create safe spaces from which to escape the persecution of an intolerant and unaccepting society. In the MCU, Inhumans seek sanctuary in a hidden community called Afterlife, where they can live in peace, shielded from humanity's prejudice, fear, and mistrust.[102] Afterlife is overseen by Jiaying, an Inhuman who is later revealed to be Johnson's mother. On living at Afterlife, Campbell tells Johnson, "Our gifts don't have to be terrifying. They're a part of us. I felt lost before I came here, too. Looking for answers in all the wrong places. But we're connected to something bigger and older than we could have ever imagined, something extraordinary. Don't walk away from it."[103]

Jiaying's backstory is crucial to our understanding of bureaucratic violence and of the subsequent encounters between S.H.I.E.L.D. and the Inhumans, for she embodies the cunning and the cruelties of humans toward

Jiaying (Dichen Lachman) provides sanctuary for Inhumans in *Agents of S.H.I.E.L.D.* (2015).

other-than-human and more-than-human beings. During the Second World War, HYDRA operative Werner Reinhardt is seeking out persons with enhanced power. At the time, Reinhardt was unaware of the existence of Inhumans, but he did suspect that certain individuals were somehow transformed when they came into contact with an object known as a "Diviner." Toward the end of the war, HYDRA agents capture several villagers in a province in China and Reinhardt exposes them to the object. All of the villagers perish, except Jiaying. Immediately, Reinhardt orders the young woman to be prepared for surgery, for, in the words of Reinhardt, "Discovery requires experimentation."[104] Allied troops, however, are fast approaching; Reinhardt is captured and sent to prison. In the process, Jiaying is freed from HYDRA and returns to her village. At some point, she undergoes Terrigenesis and her Inhuman regenerative powers are revealed. Over the years, Jiaying becomes a trusted elder in her village, helping both humans and Inhumans.[105] Jiaying falls in love and marries Calvin Zabo, a human, and subsequently gives birth to Daisy.[106]

In 1989, a group of HYDRA agents raid the village and abduct several people, including Jiaying. They are transported to Austria where Jiaying is horrified to discover that Reinhardt—now going by the name Daniel Whitehall—is alive and continuing his human experiments. Reinhardt is similarly surprised to find that the young woman he encountered four decades earlier has hardly aged.

46 CHAPTER TWO

Determined to find the source of her healing properties, Reinhardt dissects her alive. When she finally dies on the operating table, Reinhardt orders her body to be disposed. Zabo later finds the mutilated corpse of his wife and stiches her together. Jiaying's regenerative powers are then able to fully bring her back to life. However, when Zabo and Jiaying eventually make their way back to China, their daughter—Johnson—was nowhere to be found. Zabo would later relay Jiaying's story to Johnson. "They took her," Zabo coldly explains, "and said she was dangerous." He continues: "I knew better. Her gift wasn't like that. I left you with people that I trusted. I tracked your mother to Europe. But by the time I caught up with them, it was too late. Whitehall cut her to pieces. He took her organs, her blood. . . . and when he finished butchering her, he dumped what was left in a ditch, like she was garbage."[107]

To S.H.I.E.L.D., Inhumans are just another monstrous threat that requires administrative oversight. Repeatedly, Johnson pushes back against what she perceives as bureaucratic overreach, but is troubled by her own self-doubts and worries that perhaps she is a "monster."[108] Perhaps it was for the betterment of *humanity* that Inhumans like herself be subjected to the Index. Maintained by S.H.I.E.L.D., the Index is a list of "people and objects with powers."[109] When an enhanced being is made known, a team of agents assesses the individual and generates an "Index Asset Evaluation and Intake" report. The enhanced person is detained and required to undergo an extensive psychological evaluation and threat assessment.[110]

Throughout the series, the Index functions as a government-sanctioned watch list of beings deemed more-than-human and other-than-human. Once placed on the Index, Inhumans are subject to continual surveillance, monitoring, and possible detainment. In some instances, the Indexed person is deemed too dangerous and killed or, in the bureaucratic euphemism of the agency, simply "crossed off."[111] Effectively, the Index is an instrument of preemptive policing. Those who are placed on the list are prejudged as "dangerous" monsters and routinely viewed with suspicion. As such, Johnson is forced to repeatedly defend her *humanity*, decrying, "We're not monsters. You can't just lock us up in some ice prison because you don't know what else to do."[112]

Motivated in part because of his friendship with Johnson, Coulson is troubled by placing Inhumans on the Index. Coulson doesn't consider Inhumans an existential threat. "Most of them," he explains to General Talbot, "are regular people with lives and families. Who just want to be left alone."[113] However, Coulson consoles himself—or, perhaps, deceives himself—by believing that the Index is the right thing to do. Downplaying the fact that the Index codifies a particular circulation of norms, Coulson fails to recognize that

Phil Coulson (Clark Gregg), Daisy Johnson (Chloe Bennet), and Alphonso Mackenzie (Henry Simmons) routinely confront the prejudices of society in *Agents of S.H.I.E.L.D.* (2013–2020).

his actions contribute to the bureaucratic making of monsters. Conversely, Johnson—as a former hacker and an Inhuman subject to being placed on the Index—exhibits a decidedly more intimate understanding and appraisal of what's at stake. In the show's first season, Johnson learns about S.H.I.E.L.D.'s efforts to "contain" more-than-human and other-than-human threats. When S.H.I.E.L.D. agents apprehend Mike Peterson, an "ordinary" human who was injected with a variant of the super soldier serum, Coulson explains to Johnson: "We're not the only one's interested in people with powers. We'd like to contain him, yeah. The next guy will want to exploit him. And the guy after that will want to dissect him."[114] However, when Johnson transforms into her Inhuman self, Coulson's moral code is challenged. By this point he has become a close friend of Johnson's, serving both as mentor and father figure. As he witnesses the discrimination meted out against Inhumans, Coulson expresses both doubt and regret. Although firm in his conviction to defend humanity from more-than-human and other-than-human threats, Coulson also appears to hold a reverence for life toward the Inhumans.

How to balance these attitudes remains a recurrent theme for Coulson— and many other characters—throughout the series. Soon after her transformation, Johnson sobs, "There's something very wrong with me." Leo Fitz, a friend and fellow agent, attempts to comfort her. "You're just different now," Fitz explains. "You're just different now, and there's nothing wrong with that."[115] He admits, however, that being different makes her vulnerable to prejudice from other humans, for now she is at risk of "being locked up, studied, or who knows what else."[116] Indeed, Fitz realizes that, as a S.H.I.E.L.D. agent, he too is at fault, for it isn't simply how people interact as individuals that matters, but also how institutions commit harm through mundane bureaucratic procedures.

Unquestionably, Inhumans are more-than-human beings and embody physical and/or psychic powers that far surpass human capabilities. Compared to unarmed humans, Inhumans *are* potentially dangerous. In Season Four, Mace, Coulson's replacement as Director of S.H.I.E.L.D., concedes that Inhumans constitute a dangerous class of beings. Prior to a mission, for example, Daisy explains that she doesn't carry weapons, to which Mace responds: "You are the weapon."[117] It is hardly surprising, therefore, that many humans attempt to weaponize the Inhumans, including a somewhat reluctant Coulson. Talbot, conversely, strongly favors efforts to weaponize Inhumans. Although distrustful of Inhumans, Talbot recognizes the military potential of an army of Inhumans. Indeed, the "recruitment" of Inhumans is more efficient than trying to produce super soldiers, like Captain America. With Inhumans, "nature" has already done the work. Accordingly, it is only prudent to control, monitor, and weaponize Inhumans, for as Talbot explains: "We're in the middle of an arms race. World War III is going to be decided by whoever has more enhanced assets."[118] Talbot's argument, in fact, mirrors that of Fury and, arguably, the entire premise of Marvel's narrative arc. With the Avengers Initiative, Nick Fury assembled—and weaponized—a team of more-than-human and other-than-human beings in the service of humanity: the Avengers.

In *Agents of S.H.I.E.L.D.*, Coulson follows in the footsteps of his predecessor and mentor when he authorizes the use of Inhumans to serve as S.H.I.E.L.D. agents. Coulson's actions are controversial and he personally struggles with the implications of his decision. On the one hand, Coulson acknowledges—as did Fury—that nonenhanced humans are extremely vulnerable in a posthuman universe; this fact alone poses a mortal risk to humanity. On the other hand, Coulson is sensitive to the ethical consequences of using other species-beings as instruments in the service of humanity. Not surprisingly, divisions erupt within S.H.I.E.L.D. as officials differ on how to respond to and make use of Inhumans. One faction, led by Robert Gonzales, a high-ranking S.H.I.E.L.D. operative, is fundamentally suspicious and distrustful of Inhumans and views them as potential enemies; the other, led by Coulson, agrees that the Inhumans should be monitored but disagrees that more proactive measures are necessary.

Factions among the human members of S.H.I.E.L.D. come to a head when the location of Afterlife is found. Armed with this information, Gonzales sees this as an opportunity to launch a preemptive strike against the Inhumans. Johnson pleads with Coulson that the Inhumans at Afterlife should be left alone. Coulson is sympathetic, but still believes that the agency's bureaucratic oversight is for the betterment of all. Soon, Coulson acquiesces to Gonzales's plan to use Johnson as leverage against her mother. With few options, Johnson agrees to go to Afterlife first and lay the foundation for formal discussions. She is stunned, then, when S.H.I.E.L.D. arrives in full force, led not by Coulson but by Gonzales. Jiaying, subjected to human cruelties her entire life, is understandable guarded.

On meeting Jiaying, Gonzales explains that he comes in peace. "I would like to meet your people," he tells the Inhuman, and to "learn about their powers." Of any potential oversight, Gonzales simply says, "We'll keep a record and if any [humans] ever try to do someone harm, we'll be there to stop them. That's why S.H.I.E.L.D. exists."[119] The Inhumans, as Samira Nadkarni explains, are "effectively positioned as the equivalent of paroled criminals without having committed any crime other than existing outside of US control and culturally influenced perceptions of normativity."[120] From such a position, any agreement brokered with S.H.I.E.L.D. would be tantamount to giving up their freedoms. Undeterred, however, Gonzales crudely tries to make a connection with his Inhuman counterpart. Rather than seeing Jiaying as a person, he directs attention to their apparent physical "defects", comparing his scars obtained in combat with the scars that cover Jiaying's body, scars born of cruel and horrific experimentations. Gonzales offers also a necklace as a gift. Jiaying recognizes instantly the necklace, for it was the one she intended to give to Daisy years earlier, before she was tortured and butchered by Reinhardt, the necklace stolen by Reinhardt. Jiaying's anger bursts forth—her anger at Gonzales, at S.H.I.E.L.D., at humanity: "How dare you compare your scars to mine? You're nothing like me, like us. Whitehall cut me to pieces. He ripped out my organs and stuffed them in jars! I will never let that happen to my daughter. To any of my people."[121]

Determined to protect Afterlife from humans, Jiaying exposes the S.H.I.E.L.D. agent to Terrigen Crystals and he promptly turns to stone. Then, taking the gun from his lifeless hand, Jiaying shoots herself in the shoulder to make it appear as if Gonzales had first attacked her, and she killed him in self-defense. Jiaying is fully prepared for war, declaring that "S.H.I.E.L.D. found us. No matter where we go or what we do, they will hunt us down. If we are to survive, S.H.I.E.L.D. must be destroyed."[122] Jiaying is not wrong. As Nadkarni describes, the Inhumans are "either forced into a war for their independence on the one hand (with the likelihood of a high death toll) or treated as largely sub-human and 'tagged' by a secret organization. . . . whose individuals have already displayed markedly xenophobic tendencies towards the Inhumans."[123] Consequently, everything Jiaying does, she explains, is for the protection of the Inhumans. Johnson, though, is torn between her loyalties to S.H.I.E.L.D., her identity as an Inhuman, and her feelings toward her mother. "You killed Gonzales," Johnson replies. "You want a war." No, Jiaying counters: "War was inevitable. I struck first so we would have the advantage."[124]

For Jiaying, the history of humanity's species-supremacy was plain enough to see. Indeed, she had suffered for years, literally, at the hands of humanity, as scientists cut flesh and organs from her body in an attempt to discern her life-regenerating powers. For Jiaying, HYDRA and S.H.I.E.L.D. were indistinguishable: Both represented an existential threat not only to her, but also to all Inhumans. "This war started decades ago," Jiaying tells her daughter, "when S.H.I.E.L.D. was founded to guard the world against people like us, and it will never end."[125] Accordingly, Jiaying vows to seek out Inhumans everywhere "and build them a better world, where they're not hunted, not afraid."[126]

Working with S.H.I.E.L.D., Johnson and her father attempt to stop Jiaying. She does not want to oppose her mother. However, years of mistrust impede any mutual openness. Ultimately, Jiaying dies, killed by her husband. In the aftermath of the struggle, however, a Quinjet (aircraft) loaded with Terrigen Crystals crashes into the ocean and exposes various sea creatures to Terrigen mist. Subsequently, contaminated fish are processed into fish oil supplements. "Ordinary" humans who ingest the Terrigen-tainted pills are unaffected; however, those humans who carry the mutated gene undergo Terrigenesis and their enhanced powers became manifest. This marks the beginning of the so-called "Inhuman Outbreak."

Not unlike the plague, the Inhuman Outbreak is portrayed as a pandemic and widespread panic ensues as humanity is suddenly confronted with an enemy within. This is crucial, for it serves as an inflection point for the extension of bureaucratic power throughout society. In the Middle Ages, Foucault writes, the "plague appears as the moment of panic and confusion," and yet the "plague also brings the political dream of an exhaustive, unobstructed power that is completely transparent to its object and exercised to the full."[127] Foucault elaborates that efforts to contain outbreaks resulted in the "spatial partitioning and control of plague-infested towns."[128] The purpose was not to exclude diseased individuals, but was instead a matter "of establishing and fixing them, of giving them their own place," thereby limiting the contamination of their presence.[129] This corresponds with what Michel Foucault terms the power of "normalization."[130]

In his discussion of the "dangerous individual," Foucault details how this figure is made abnormal and thus threatening to society. In other words, the monstrous Other deviates from and thus threatens societal norms. For Foucault, "the norm, consequently, lays claim to power" in that "the norm's function is not to exclude and reject. Rather, it is always linked to a positive technique of intervention and transformation, to a sort of normative project."[131] In Marvel's fictional world, the uncertainties surrounding the presence of other-than-human beings—the Inhumans—creates the conditions that intensify the ongoing penetration of bureaucratic control throughout society. Humanity is forever the norm; those who deviate from it are forever abnormal. As such, the unleashing of bureaucratic power toward Inhumans is a way of including the monstrous through exclusion, that is, to circumscribe the moral standing of other-than-human beings.

As the existence of Inhumans becomes more widely known throughout the world, Mack concedes that humans are "terrified." However, Mack also understands how frightened humans will respond. Mack explains that terror makes them—*humans*—"dangerous." In this revelatory moment, Mack makes clear that humans become monstrous in their prejudices toward the Inhuman Other and this explains why the government initiated "a witch hunt for powered people."[132] Mack's words are prophetic and innocent Inhumans are made scapegoats. As US senator Christian Ward remarks, "The American people are looking for a simple enemy. . . . It's what makes them feel safe."[133] Indeed, sensationalist media reports, backed by government officials, whip the public into a frenzy of fear and anger directed toward the Inhumans. At one point, Coulson quips, "We might as well start handing out pitchforks and torches."[134] Later, decrying the rise in violence against Inhumans, Elena "Yoyo" Rodriguez—herself an Inhuman—laments, "So many people looking for an excuse to hate."[135]

Hatred, prejudice, and violence is formally authorized when US president Matthew Ellis announces the formation of a "special task force to neutralize these alien threats on our own soil—the Advanced Threat Containment Unit" or ATCU. According to the president, the ATCU. "will be given full license to act with whatever authority is necessary." And to be perfectly clear, the president insists, "Any threat will be eliminated."[136] Interestingly, Ellis has a somewhat checkered history with more-than-human and other-than-human beings. Early in his presidency, for example, Ellis banned biotech research, which he considered immoral. However, Ellis also supported the development and deployment of exoskeletons based on Tony Stark's Iron Man suit in the US armed forces, notably his promotion of Colonel James Rhodes (War Machine) to be rebranded as "Iron Patriot."[137] It appears as if Ellis supports technologies that enhance human capabilities, such as exoskeletons, but balks when technologies appear "unnatural," that is, go against human nature. It follows, therefore, that Ellis would not only view the Inhuman Outbreak as a threat, but would also consider the Inhumans as monstrous. From the highest echelons of government down, Inhumans—because they were seen as being abjectly different—were declared enemies of the state, subject to mass incarceration, if not extermination.

With S.H.I.E.L.D. having been previously infiltrated by HYDRA, Ellis premised that the ATCU would become its successor. Ellis appoints Rosalind Price, a no-nonsense CIA operative, as director of the agency. Formally, the ACTU is mandated to contain and incarcerate Inhumans. Agents are, however, sanctioned to use lethal force and, at times, it appears as if this is the desired first option.[138] On taking the helm of the ACTU, for example, Price publicly declares that "These altered humans are a threat and I'm in charge of neutralizing that threat."[139] Echoing Ellis, she affirms also that "the laws of nature have changed and until the laws of men change to reflect that, we can only do what we feel is right."[140] Markedly, since existing laws do not reflect the new posthuman reality, Price and the ACTU are prepared to use any and all means necessary to contain the supposedly "unnatural" threat. Essentially, Inhumans are made monstrous and thus positioned outside the law.

During the Inhuman Outbreak, one of the first people to undergo Terrigenesis was a construction foreman known as Joey Gutierrez. For much of his life, Gutierrez struggled as a closeted gay man, feeling alienated, isolated, and ostracized. As he later reveals to Johnson, "I've lived with a secret before. I was miserable until I came out with it."[141] In short, he already saw himself monstrous, unable to integrate fully into a human-centered heteronormative society, that is, until he decided to embrace his sexuality. He began dating a man, publicly, and went to the gym to get in better shape. It was in the course of his new-found health regime that he took fish oil pills that, unbeknownst to him, were contaminated with Terrigen crystals. After undergoing Terrigenesis, Gutierrez emerged as an Inhuman with the ability to "liquefy certain metals spontaneously at a distance of up to three meters," an ability he neither wanted nor understood how to control.[142] Predictably, Gutierrez accidentally began melting objects in his presence, causing significant damage throughout his neighborhood.

In short order, ATCU agents, including Price and fellow agent Luther Banks, a former US Marine, arrived on the scene. When Gutierrez is cornered, Price instructs her agents to "use lethal force if necessary" to subdue the threat. Terrified and confused, Gutierrez calls out, "Lethal force? I'm not trying to do this. I need help!"[143] The response of Price and her agents is telling insofar as it underscores the "shoot first and ask questions later" attitude indicative of many government agencies tasked with security enforcement throughout the MCU. Too often, societal problems, such as homelessness, drug abuse, and mental health concerns, are met with legally sanctioned force enacted by police departments; the fictional ATCU is no exception. However, in this instance, agents of S.H.I.E.L.D. arrive on time and are able to safely extract Gutierrez without any further damage. Later, Johnson explains to a frightened and bewildered Gutierrez what he has undergone. "This chemical compound, called Terrigen," she tells Gutierrez, "was recently released into our ecosystem. And you, Joey, are one of the first people with this gene to come into contact with it." Not fully comprehending the nature of his transformation, Gutierrez asks if he is an alien. "Part alien," Johnson responds, "welcome to the club; we call ourselves Inhumans."[144]

Publicly, the US government purports to care about Inhumans and is actively working to help them assimilate into "normal" society. In practice, however, the ACTU is warehousing—or incarcerating—Inhumans kept in suspended animation at a facility known as Endotex Labs. None have been tried or convicted; indeed, none were even charged with criminal activity. Their unlawful detainment is instead framed as a health crisis, their imprisonment necessary to protect humanity until a *cure* is found for the Inhuman disease. Price is fully aware of ongoing abuses but only belatedly recognizes the inhumane treatment of Inhumans for what it is. In the end, Price understands also that she has become a monster. "I find it hard to keep my humanity in all this," she reveals to Coulson.[145]

Unlike Price, Coulson believes it possible for Inhumans to assimilate back into society; however, he also remains committed to using Inhumans as secret agents in defense of humanity. As such, with the assistance of Andrew Garner, a psychologist who has considerable experience with S.H.I.E.L.D., Coulson initiates a program codenamed Caterpillars.[146] In a way, the Caterpillars program parallels the Avengers Initiative: Both bring together a select team of enhanced beings to counter threats against humanity. However, Coulson envisions the program also as providing a temporary sanctuary for Inhumans, not unlike Afterlife.

Johnson is asked to oversee the Caterpillar program. Being an Inhuman and a S.H.I.E.L.D. operative, Johnson is acutely sensitive to the exploitation of the more-than-human and other-than-human beings. However, she also sees the Caterpillars program as an opportunity for Inhumans and other enhanced persons to demonstrate their commitment to defending humanity. Here, her attitude reflects that of many marginalized populations: to position themselves as hyperpatriotic and hyperloyal to the state. Johnson is not unaware of the challenges involved, but sees few alternatives. The ATCU was actively hunting, detaining, and even eliminating Inhumans; and yet, the world also confronted multiple threats from supervillains that humans could not fight on their own. "My mother was right about one thing," Johnson tells Coulson. "People like me need to be kept a secret. Not like the Avengers, out in the open. If we do this, we need to be—" Coulson interjects: "Anonymous. That's the idea. But it's not if we do this. We have to do this. We don't have a choice."[147] Johnson later justifies her participation in the program to Garner: "My mother created a halfway house for Inhumans, and it was not enough. I want Inhumans taking action with S.H.I.E.L.D. to see that being different can mean making a difference."[148]

The Caterpillars program was only partially successful. Perhaps not unexpectedly, Johnson encounters challenges in recruiting other Inhumans. Gutierrez, for example, trained with Johnson, only later to quit the program. On the whole, Inhumans, including Gutierrez, simply wanted to live their lives free from the prejudices of humans and free from government witch hunts. However, many Inhumans expressed also their reservations about working for a secret government organization. A notable exception is the appearance of Elena "Yo-Yo" Rodriguez, an Inhuman who eventually becomes a prominent member of S.H.I.E.L.D. Markedly, the character of Rodriguez serves an important function throughout the show in that audiences see through her eyes the subtle and not-to-subtle forms of discrimination and doubt harbored by otherwise sympathetic humans. In fact, Yo-Yo also personifies the frustrations experienced by Johnson. Seemingly at every step throughout the ill-fated program, Johnson and her team of more-than-human beings is met with suspicion and skepticism—including from other "human" S.H.I.E.L.D. agents.

HUNTING MONSTERS

The world needs to know that Inhumans are a disease,
and we're the only cure.
—ANONYMOUS WATCHDOG MEMBER *(AGENTS OF S.H.I.E.L.D.*, "EMANCIPATION")

As the persecution of Inhumans continues, Coulson expresses even deeper reservations about maintaining the Index. In a particularly revealing conversation with Talbot, Coulson begins: "If you force them to register. . . ." Talbot interrupts, explaining that "it's a protected list, it's highly classified. You had an index." "Which is why," Coulson counters, "I know nothing ever good comes from putting people on lists. Eventually they get misused or fall into the wrong hands. And then innocent people suffer."[149] Coulson's concern will prove prophetic. In season three of *Agents of S.H.I.E.L.D.,* a new threat to the Inhumans emerges—an armed militia of humans known as the Watchdogs. As self-styled vigilantes, members believe that the government—including its secretive apparatuses—are not doing enough to protect humanity from the Inhumans. Indeed, as one Watchdog explains, "We don't need registration, we need extermination."[150] Thus, while S.H.I.E.L.D. works to place Inhumans on the Index, self-styled vigilantes begin harassing Inhumans, including committing acts of mortal violence.

The theme of vigilantism permeates the MCU. The *Guardians of the Galaxy,* for example constitute a multi-species group of more-than-human beings who, for different reasons, literally guard the galaxy against dangerous beings. However, the theme is addressed most directly in Marvel's many Netflix-produced series, such as *Daredevil, The Punisher, Jessica Jones,* and *Luke Cage.* Unlike the blockbuster films of the MCU, these shows "were uniquely pitched as critical commentaries on social issues in the American polity."[151] *The Punisher,* for example, examines the traumas of war, loss, and toxic masculinity, while *Jessica Jones* and *Luke Cage* explore themes of gender-based violence and systemic racism, respectively.[152] *Daredevil,* along with the show's protagonist, Matt Murdock, is notable in that it eschews simple dichotomies of "good" and "evil." Here, Murdock "is self-avowedly a classic vigilante in the sense of operating outside the constraints of societal criminal law."[153] As such, the villains of the series "tend to demonstrate a humanity worthy of some respect, and thus worthy of trying to salvage, which plays into Murdock's sense of rehabilitative justice."[154]

In *Agents of S.H.I.E.L.D.,* the Watchdogs unambiguously constitute a hate group. Motivated by their fear of Inhumans, literally, dehumanized Others, Watchdog members exhibit extremist attitudes of human exceptionalism and human supremacy. There is no effort to demonstrate any reverence for life of the Inhumans; as the monstrous Other, Inhumans are always and already abject, forever outside of the Watchdog's moral community. As Tucker, a member of the Watchdogs, explains, his desire is nothing less than to "eliminate the Inhuman[s]."[155] Similarly, in the words of another, unnamed member, it was necessary for humans to "kill the aliens and take back this planet."[156]

As the Inhuman "outbreak" spreads, the Watchdogs believe that the government is either unable or, perhaps more conspiratorially, unwilling to eliminate the more-than-human threat. As one member, known only as Watchdog Alpha, explains, "The government is keeping secrets, things we

In *Agents of S.H.I.E.L.D.* (2016), members of the human supremacist hate group, the Watchdogs, want to exterminate all Inhumans.

don't know until cities fall out of the sky. The Avengers, more like them every day, protected, hidden. Not anymore! We want this information released, a list of who they are, where they are or you will have war on your hands. We are the Watchdogs. You will obey."[157] Nothing less than a self-styled civil war against those perceived as outside the norm would suffice.

For other humans, such as Anton Ivanov, a Russian crime boss, hatred toward the Inhumans stems from a different notion of human supremacy. Ivanov is driven "to eliminate the Inhuman problem" not so much because they are considered "dangerous" and thus pose an existential threat to humanity.[158] Indeed, he can respect to a degree the more-than-human abilities of Inhumans. Instead, Ivanov's disgust centers on his view of Inhumans as being evolutionary cheaters. Ivanov explains, "These Inhumans. They did not suffer. . . . labor to become what they are. They have great power but they didn't earn it. They are unnatural things."[159] And of Johnson specifically, Ivanov accuses: "You're a genetic cheater."[160] More broadly, Ivanov extends his hostility toward S.H.I.E.L.D., accusing the agency of providing a "safe haven for Inhumanity."[161] Monstrous in his own violence toward Inhumans, Ivanov serves as a warning to a posthuman future populated by genetically engineered people.

The scapegoating and persecution of Inhumans is deftly manipulated by government officials and media commentators, who stoke the fires with violent rhetoric. For example, Felix Blake, a renegade and embittered S.H.I.E.L.D. agent, upholds, "Hate's a good motivator."[162] And what is the source of Blake's hatred toward the monstrous Other? In part, Blake seems

Ellen Nadeer (Parminder Nagra), a US senator, covertly forms the Humans First Movement in *Agents of S.H.I.E.L.D.* (2016).

to speak for many ordinary humans. In a tense confrontation with Phil Coulson, Blake asks, "What has S.H.I.E.L.D. done in the name of protection? We brought the Chitauri to Earth."[163] In other words, Blake accuses the government as the source of humanity's problems. In defense of his violent activities toward enhanced individuals, consequently, Blake claims that he is just "trying to make the world a safer place."[164]

Whereas the Watchdogs represent a grassroots movement, anti-Inhuman prejudice and human exceptionalism is present at all levels of government. Among the most prominent is Ellen Nadeer. Nadeer is a United States senator and is outspoken in her antipathy toward Inhumans and other enhanced individuals. Nadeer describes Inhumans as "alien filth" and frames the Inhumans as a disease infecting humanity—an "Inhuman epidemic."[165] For a human supremacist such as Nadeer, even the word Inhuman is "frightening."[166]

Similar to the Watchdogs, Nadeer believes that Inhumans pose a threat to national security and must be eradicated completely. "We can no longer pretend like the Inhuman threat doesn't exist," Nadeer explains, for "each and every one of them is a lethal weapon and now those weapons are aimed directly at us. Unless we take a stand and fight back now while we can, we will be complicit in our own extinction."[167] Believing Inhumans to be nothing more than "lethal weapons," Nadeer does not want to contain and control Inhumans but instead hopes to eliminate them altogether. During a particularly tense confrontation, Johnson calls Nadeer an "anti-Inhuman hate-monger" and decries that "the world has become a very dangerous place because of people like you."[168] Inhumans, Johnson continues, "are more likely to be victims of hate crimes, which is not fair. They are our friends, they are our co-workers, our brothers. They all deserve help, wherever they may be."[169] Nadeer is unmoved, responding simply that "It's people like me who make the world safe," people who are willing "to do the right thing."[170]

It is important to note that Nadeer's antipathy toward Inhumans and other enhanced individuals stems from her traumatic experiences of the Battle of New York, a signature event that reverberates throughout the MCU. As Maria Hill reflects, "The Battle of New York was the end of the world. This—now—is the new world. People are different." She expounds, "Everything's changing. A little while ago, most people went to bed thinking that the craziest thing in the world was a billionaire in a flying metal suit. Then aliens invade New York that were beaten back by among others, a giant green monster, a costumed hero from the '40s, and a God."[171] During the Battle of New York, which pitted the Avengers against invading Chitauri aliens, Nadeer's mother was killed. Tormented by the event, Nadeer develops and harbors a deep-seated hatred of any other-than-human or more-than-human being. "You and I were both there," she reminds her brother, Vijay, "when the Chitauri killed mom, aliens invading our world. Everything changed that day."[172] Nadeer's brother, however, is an Inhuman, a plot twist that troubles her attitudes toward more-than-human and other-than-human beings. Unable to cultivate empathy for her brother, who had no control over his transformation, Nadeer ultimately kills Vijay. Nadeer's hatred drives her to form the Humans First Movement. If, following Albert Schweitzer, "ethics are responsibility without limit towards all that lives," we find in Nadeer and those who support her, the embodiment of immoral actions.[173]

Ideologically, the Humans First Movement is species supremacist in orientation and thus provides an effective outlet to tap into humanity's collective fears. As the dystopian scholar Gregory Claeys writes, "power lies in numbers, and in the vaunted fictitious unity they exhibit." The Humans First Movement, in this sense, calls upon a discredited fiction of universal humanism, sidestepping the historical exclusion of minority populations. Markedly, Claeys continues, the larger the crowd, the more powerful the sense of equality and, equally, of loss of self, of release of anxiety, of the abrogation of responsibility, and of the feeling of immortality in the life of the species group.[174] Here, we see how the Humans First Movement replaces the uncertainties and anxieties of facing a posthuman future with a sense of belonging. In fact, one might argue that outdated notions of humanism offer comfort to people long marginalized and dehumanized. Members of the Humans First Movement, for example, are seemingly representative of all categorizations of humanity although, certainly, the inclusivity of the movement is more implied than confirmed. The main takeaway, though, is that many humans—faced with the existential crises that humans are not the center of the universe or do not sit at the apex of God's plan—are drawn to the Humans First Movement in a monstrous act of solidarity. Group-think becomes all-powerful; and individuals within the group are able to engage in activities, including collective violence, in ways otherwise unimaginable.

When members of hate groups such as the Humans First Movement come together and do harm against those beings considered monstrous, the explanation is found not in the persecuted but instead the perpetrator. In Marvel's posthuman universe, the Inhumans (with few exceptions) exhibit no intention to harm humans; indeed, most continue to see themselves as humans, albeit with different abilities. For many "normal" humans, however, Inhumans are abject and the mere fact of their existence is enough to generate fear and anxiety. They are forever different, not normal; their presence underscores an era long-since passed and fuels efforts among the supremacists to return society to a time when the abject Other seemingly didn't exist. Indeed, throughout the narrative arc of the MCU, there is little consideration to extend moral standing to Inhumans; many humans refuse to recognize the other-than-human beings as simply wanting to live a meaningful life.

Harboring extremist views to protect humanity against nonhuman beings, Nadeer is especially reluctant to trust the government to deal effectively with the Inhuman threat. For Nadeer, S.H.I.E.L.D. is a puppet agency of the Inhumans. This is not some outlandish conspiracy. Daisy Johnson, a highly visible member of S.H.I.E.L.D., is a known Inhuman and her status as an agent does lend credence to Nadeer's suspicions. "What is S.H.I.E.L.D.'s purpose," Nadeer questions, "if it's not protecting us from the Inhumans. Why are human lives not S.H.I.E.L.D.'s priority?"[175] Any doubt is removed when she subsequently learns that S.H.I.E.L.D. is secretly harboring Inhumans around the world, in direct violation of the Sokovia Accords.

Daisy Johnson long feared the threat posed by registering and indexing other-than-humans, and the actions of Nadeer would ultimately prove her correct. Driven to exterminate the Inhumans, Nadeer forms a secret alliance with members of the Watchdogs, including the wealthy Russian industrialist, Anton Ivanov. Crucially, all signatories to the Sokovia Accords had access to the Inhuman registration list. Subsequently, Nadeer, as a government official, also had access and she was able to pass this information to the human extremist militia and assist in their hunting and killing of Inhumans. However, Nadeer assisted also in staging a worldwide coordinated attack on the Inhumans, designed to foment a human-led rebellion against the more-than-human and other-than-human beings. The strategy consisted of a coordinated false flag operation where humans, posing as Inhumans, would initiate a series of highly visible terrorist attacks.

Using advanced technologies stolen from S.H.I.E.L.D., members of the Watchdogs set off an EMP (electromagnetic pulse) device first in Miami, Florida, resulting in a massive power outage. Subsequently, a video is released to the media that appears to be an Inhuman claiming responsibility for the attack, demanding that the registration list be cancelled and warning of other attacks. In reality, the Inhuman in the video was a member of the Watchdogs posing as an Inhuman. The ruse works and many humans begin in fact to turn on the Inhumans. Further fanning the flames of hatred and prejudice toward the Inhumans, Nadeer goes public and announces that Inhumans are the source of the attacks. In addition, Nadeer casts blame on S.H.I.E.L.D. in an apparent attempt to discredit the agency. To the American public, Nadeer decries, "People are scared that Inhumans don't share our core values, and who could blame them? How many of our citizens have died under S.H.I.E.L.D.'s watch?"[176]

Following the highly visible and highly staged attacks in Miami, similar acts of violence are carried out in Washington, DC, Los Angeles, London, Moscow, Buenos Aires, and Athens. Coulson quickly recognizes the implications of the coordinated attacks. "If they're hunting Inhumans here," he concludes, "they're hunting them in other cities."[177] Equipped with this knowledge, S.H.I.E.L.D. agents are able to thwart the operation, but not before significant damage is done. In the aftermath, a Watchdog informs Nadeer that seventeen Inhumans had been killed. Nadeer responds, "Good. Good start."[178]

The violence perpetrated by militias and hate groups, such as the Watchdogs, against more-than-human and other-than-human beings often takes center stage in Marvel's cinematic universe. This is not unexpected; indeed, for the superhero genre, it is par for the course. However, *Agents of S.H.I.E.L.D.* also illustrates how seemingly mundane bureaucratic procedures facilitate violence through the administrative making of the monstrous Other. So defined, Inhumans and other "extraordinary" individuals are subject to prejudice, bigotry, and hatred; government agencies are called upon to monitor, detain, and possibly eliminate those beings who exist outside of societal norms. Of course, if the government is unable, or unwilling, to perform these functions, private militias and vigilantes are ready to step into the breach.

CONCLUSIONS

When Stan Lee first introduced a species of mutant humans—the X-Men—he wanted to call attention to the persecution of marginalized peoples and to challenge the discriminatory treatment of ordinary people rendered abnormal by society. This theme remains dominant throughout the MCU, but is especially prominent in the television series *Agents of S.H.I.E.L.D.* Indeed, as the character of Mace, Coulson's replacement as Director of S.H.I.E.L.D. affirms, "I believe that Inhumans deserve the same rights as anyone else in this country."[179] Surprisingly, though, the show has received minimal critical commentary—perhaps because it is no longer considered canon. Also, when a critical eye is cast toward the series, it most often focuses on the undercurrents of US militarism and imperialism that permeated especially the first season. According to McSweeney, for example, "the show rarely pauses to reflect on the nature of such a powerful and clandestine organization" and, as such, "reifies the role of powerful and secretive government agencies in the real world that we are asked to trust unequivocally; secure in the knowledge that they are acting in our own interests."[180] Here, McSweeney's criticism is appropriate but, I suggest, there is more to be seen. As Darren Franich finds, "although generally portrayed as a good-guy super-organization, S.H.I.E.L.D. has some dark moments, and many of the most interesting stories about the group represent contemporary concerns about government overreach."[181] This extends notably into the bureaucratic treatment of those people who exist outside the norm. There is, in other words, a deeper, more ethical component raised by the series, one that calls into question the treatment of more-than-human and other-than-human beings in a rapidly transforming posthuman world.

In the introductory chapter, I suggested that Albert Schweitzer's framework is useful in helping us think through humanity's obligation to more-than-human and other-than-human beings. In *Agents of S.H.I.E.L.D.*, the treatment of Inhumans provides a cautionary tale of our uncertain posthuman future and of humanity's (in)ability to recognize their will-to-live. Whereas S.H.I.E.L.D. agents such as Coulson, Mace, Mackenzie, Fitz, and Simmons are largely sympathetic toward enhanced individuals and question the morality of registering, monitoring, and detaining more-than-human and other-than-human beings, other agents appear more callous in their conduct. These struggles are also manifest with the introduction of other government agencies in the MCU.

Notable here is the US Department of Damage Control (DODC).[182] When the DODC makes its first appearance in the MCU, in the film *Spider-Man: Homecoming* (2017), its original mission was to "oversee collection and

Agent P. Cleary (Arian Moayed) is determined to defend Earth from other-than-human beings.

storage of alien and other exotic materials" following the Battle of New York.[183] Created in partnership with Stark Industries, the DODC was initially a subdivision of S.H.I.E.L.D. With the collapse of S.H.I.E.L.D., however, the DODC assumed many of the functions once performed by the former agency. This included the identification, detainment, and interrogation of suspected enhanced individuals and their possible incarceration at a maximum-security center known as the Damage Control Supermax Prison. Early on, ABC Television planned a series centered on the Department of Damage Control, but eventually the project was scrapped.[184] Still, the bureaucracy has appeared in several films and streaming series, including *Spider-Man: No Way Home* (2021) and *Ms. Marvel* (2022).

To date, not much is known about the personnel who comprise the DODC, although two characters stand out: Agents Sadie Deever and P. Cleary. The former makes her first appearance in *Ms. Marvel* and is portrayed as an agent ruthless in her pursuit of more-than-human and other-than-human beings, including the mutant Kamala Khan (Ms. Marvel). Indeed, in one episode Deever disobeys orders from her superiors and authorizes an attack on a high school where a group of suspected enhanced individuals, including Ms. Marvel, are hiding. Thus far, the source of Deever's antipathy toward mutants is unclear and there is no indication that Deever's storyline will gain prominence in the MCU. A second character, P. Cleary, first appears in *Spider-Man: No Way Home* and returns in *Ms. Marvel*. As a DODC agent, Cleary is tasked with monitoring and potentially detaining, interrogating, and possibly arresting more-than-human and other-than-human beings, including both Peter Parker (Spider-Man) and Ms. Marvel. Similar to Deever, Cleary appears also to be highly motivated in stopping enhanced individuals who are believed to be operating outside the law, although there is no indication as yet that he is as ruthless as Deever. Moving forward, the human agents of the DODC will no doubt provide insight into the ethics of so-called ordinary humans who inhabit a rapidly changing posthuman future.

Chapter Three

MARVEL'S MODERN PROMETHEUS

The line between scientists and mad scientists is paper thin.
—ALPHONSO "MACK" MACKENZIE *(AGENTS OF S.H.I.E.L.D.,*
SEASON 4, EPISODE 14, "THE MAN BEHIND THE SHIELD")

In Mary Shelly's *Frankenstein*, the titular character, Victor Frankenstein, brings to life a monstrous being, known only as "the creature." The full title of Shelly's classic is *Frankenstein, or The Modern Prometheus*. In Greek mythology, Prometheus is the creator and protector of humankind; as a paternal figure, Prometheus imparts both knowledge and technology to humans. Prometheus, however, angers Zeus when he steals fire to give to humanity, and is punished for this transgression. The modern Prometheus, as conceived by Shelley, is "the unthinking creator who fails, whether intentionally or unconsciously, to be responsible for his creation, thereby creating evil."[1]

Since its publication in 1818, Shelly's novel has sparked considerable debate on the perils of humanity's hubris in tampering with nature. As Lester Friedman and Allison Kavey explain, Frankenstein's monster neatly corresponds with the most significant questions of Shelley's time. The figure of Frankenstein, for example, embodies the central question of how far humans should seek to mimic the powers of God or nature in the name of scientific investigation. The tragedy of Frankenstein's monster, Friedman and Kavey suggest, "is that he desires acceptance by a group frightened by his very existence, horrified at his physical appearance, and bent on his destruction. Such vicious rejections result in the Monster exacting a terrible revenge, thus validating the aggressively hostile vision that initially inspired his exclusion."[2]

The specter of Frankenstein's monster haunts humanity's ongoing attempt to create artificially intelligent robots and remains one of the most popular and well-known tropes in science fiction.[3] In 1939, two brothers, Otto and Earl Binder began a series of adventure tales for the pulp magazine *Amazing Stories*. Under the pen-name of Eando, the brothers' fictional narrative revolve around a robot, Adam Link, and his wife Eve, who have to overcome fear and distrust from humans to fit into human society. In the first installment, "I, Robot," Adam is accused of murdering his creator and stands trial, effectively defending himself but also robotic beings in general. Isaac Asimov would later appropriate the title "I, Robot" for his own collection of robotic tales. Notably, Asimov would also coin the term "Frankenstein Complex" to describe the fear of losing control of our own mechanical, technological creations.[4]

The Frankenstein Complex features prominently in Marvel's cinematic universe. In *Avengers: Age of Ultron* (2015), for example, the one-time weapons manufacturer Tony Stark creates an artificial intelligence system known as Ultron, and in *Agents of S.H.I.E.L.D.*, the transhumanist Holden Radcliffe builds an android named AIDA. In both instances, not unlike Shelley's cautionary tale, Marvel's artificial beings threaten humanity. Beyond the spectacle of violent human-robot encounters, however, Marvel's renditions tap into the perennial question of what it means to be human. When we see Frankenstein in the context of robots, Michael Szollosy writes, we realize we fear more than technology or technology gaining autonomy and moving beyond our control. Instead, Szollosy suggests, we learn that what we fear is the very quality of ourselves that enables us to create the monster. Certainly, there is an element of ego involved—Stark's narcissism is legendary in the MCU—but the creation of artificial Others serves as a warning that "we are becoming the robots that we so fear." And to that end, Szollosy explains, "the re-imaginings of the Frankenstein myth in robotic monsters are not merely suggesting that it is the hubris, or forbidden knowledge, or the Evil Scientist that we must fear, but something more nuanced and slightly more complex."[5]

Our efforts to better design artificial intelligence have turned our gaze inward. When we think critically about the various human-robot encounters portrayed in science fiction and the superhero genre, including those depicted in the MCU, most contain an implicit anthropology: an understanding of what it means to be human.[6] These fictional narratives, in other words, purport to teach us something about ourselves, about how we think, and how we should relate to nonhumans.[7] The MCU, for example, is populated with countless technological beings, some benign, others malevolent, and still others occupying more ambiguous positions that compound the uncertainty—and fear—of their human counterparts. As such, the MCU provides an opportunity to reflect on the design, manufacture, and presence of these beings and how their existence mediates the meaning of being human. In the real world, we are standing on the precipice of a fundamental challenge to moral thinking, precipitated in part by our own artificial creations. How we respond will, in turn, have a profound effect on how we understand ourselves and our relation to more-than-human and other-than-human beings in an uncertain posthuman universe.

ARTIFICIAL BEINGS

The robot apocalypse is finally here.
—ALPHONSO "MACK" MACKENZIE *(AGENTS OF S.H.I.E.L.D.,*
SEASON 4, EPISODE 9, "BROKEN PROMISES")

When S.H.I.E.L.D. agent Alphonso "Mack" Mackenzie bemoans the arrival of the robot apocalypse, he conveys an existential dread experienced by many viewers: a palpable fear that artificial intelligence systems may portend the end of humanity. Indeed, as Russell Blackford writes, "it is well ingrained in science fiction's mega-text that androids, robots, suitably advanced computers, Artificial Intelligences, and similar beings will probably rebel and destroy their makers."[8] And yet, as several pundits observe, the robot "invasion" has already begun. David Gunkel, for example, declares that the robot invasion "is an already occurring event with machines of various configurations and capabilities coming to take up positions in our world through a slow but steady incursion."[9]

Writing in 2011, Adam Keiper and Ari Schulman proclaimed that "we have already entered the age of increasingly autonomous robots."[10] Presently, robots are ubiquitous in many parts of the world, making their presence known (or staying out-of-sight) in a variety of domains. In the domestic sphere, robotic machines vacuum floors, mow lawns, iron clothes, and perform myriad other household chores; in the health and medical professions, machines diagnose illnesses, assist in medical procedures, and provide therapeutic care; and on the battlefield, robots defuse bombs, engage in surveillance activities, and "eliminate" targets—including humans.[11] Many of these robots are humanoid in appearance, but most are not. Regardless, the number of robots that coexist with us is increasing exponentially; no human activity remains untouched by artificially intelligent robotic beings. And yet, paradoxically, "most people have little sense of what real robots do or look like."[12] As John Jordan explains, so conditioned are we to be on the lookout for Terminator-like robots that we too often fail to see robots in their more banal forms.

Today's artificially intelligent robots are mostly circumscribed in their preprogrammed functions. Social robots, for example, are coded and thus circumscribed to perform specialized tasks, in homes, hospitals, or some other location. A robot designed to greet guests in a hotel cannot quit and seek employment in a hospital. Indeed, even the most advanced robots we currently encounter are not completely autonomous, that is, equipped with a "mind" of their own. Far from being autonomous beings, robots remain tethered to human design, manufacture, and manipulation. What happens, though, if our mechanical creations break free and no longer accept their allotted instrumentalist role? Will a self-aware, artificially intelligent being rebel and overthrow those who programmed it to a lifetime of servitude? Before tackling these questions, however, some definitional clarity is required.

To begin: What exactly is artificial intelligence?[13] Intelligence, broadly conceived, may refer to the ability to accomplish complex goals.[14] Artificial intelligence, subsequently, can be defined as intelligence displayed or simulated by code (algorithms) or machines.[15] Even this seemingly straightforward definition, however, is fraught with unexpected problems. As Max Tegmark explains, because there are many possible goals, there are many possible types of intelligence. Consequently, when we adopt a more capacious definition of intelligence, we find that intelligence is not something uniquely or even especially "human" at all.[16] Indeed, an offshoot of decades of work on artificial intelligence has demonstrated repeatedly that humans have no monopoly on intelligence. Humans are not unique in their capacity to accomplish goals, complex or otherwise; myriad other species-beings exhibit comparable behaviors.[17] It follows, therefore, that whales and worms, dogs and dung beetles, and all the other species-beings that coexist with humans are intelligent in their own species-specific ways. And yet, in many ways, this is beside the point.

"What would it mean," James Bridle wonders, "to build artificial intelligences and other machines that were more like octopuses, more like fungi, or more like forests?"[18] In theory, we could, to Bridle's question, design any number of artificial intelligence systems based on different "forms" of intelligence. We could, for example, design robotic dogs equipped with dog-like intelligence. In practice, though, when scientists and engineers design and manufacture artificial intelligence systems the endgame is an entity that has "the ability to solve problems, learn, and take effective, human-like action, in a variety of environments."[19] In part, this effort circles back to a paramount reason *why* humans are developing artificial intelligence systems: to assist or to replace humans in a variety of settings.

Philosophers have long contemplated the existence of intelligent machines as metaphors to understand humanity; so too have writers explored the possibility of intelligent machines to reflect on being human.[20] In 1956, however, John McCarthy and Marvin Minsky hosted the Dartmouth Summer Project on Artificial Intelligence, an invited workshop funded by the Rockefeller Foundation, at Dartmouth College.[21] In the workshop proposal, McCarthy, Minsky, Nathaniel Rochester, and Claude E. Shannon explained that participants would "proceed on the basis of the conjecture that every aspect of learning or any other feature of intelligence can in principle be so precisely described that a machine can be made to simulate it." As such, the purpose would be "to find how to make machines use language, form abstractions and concepts, solve kinds of problems now reserved for humans, and improve themselves."[22]

The Dartmouth Conference was a product of its time. Held during the early years of the Cold War, the workshop followed years of rapid advances in the design, development, and application of computers and cybernetics.[23] This marked the formal incorporation of "artificial intelligence" into the scientific community and set the initial parameters of the field. Prominently, the purpose was to *simulate* human intelligence and not, as Coeckelbergh underscores, to recreate human intelligence in a machine.[24] During the ensuing two decades, remarkable achievements were made in the field of artificial intelligence. In the mid-1960s, for example, Joseph Weizenbaum created ELIZA, a computer program capable of simulating a conversation with humans; Herbert Simon, Cliff Shaw, and Allen Newell, likewise, developed a program capable of solving simple problems.[25] In this heady atmosphere, optimism prevailed and scientists, engineers, and popular commentators made repeated references that a machine as intelligent as a human being was forthcoming, possibly within a few years.[26]

The pursuit of artificial intelligence did not proceed without setbacks. Whereas the 1950s and 1960s were the halcyon days of artificial intelligence research, a palpable lack of progress—especially when seen against the backdrop of widely optimistic claims—contributed to a decline in funding and interest. Consequently, throughout the 1970s, several governments—and, more crucially, corporate sponsors—began to sour on the idea that artificial intelligence was possible. In 1973, the British mathematician James Lighthill, for example, questioned the optimistic outlook given by AI researchers, cautioning that machines would only ever reach the level of an "experienced amateur" in games such as chess.[27]

Perhaps in response to repeated critiques (and the need for continued funding), scientists and engineers focused their work mostly on the development of artificial intelligence systems equipped with discrete capabilities or designed to fulfill specific, practical tasks, that is, *artificial narrow intelligence* (ANI).[28] IBM's Deep Blue chess computer is an example of artificial narrow intelligence. Despite its impressive hardware and software, Deep Blue could *only* play chess; its success was dependent entirely on its sheer computational power rather than a deep understanding of chess.[29] If challenged to a game of checkers, Deep Blue could do nothing. Simply put, ANI's reality is limited to its predetermined capabilities; as Gonenc Gurkaynak and colleagues conclude, these computers can solve complex problems in the blink of an eye, but they do not have any preconception of things other than the information provided to them by their creators.[30] This is not to diminish the importance of artificial narrow intelligence. Indeed, the presence of ANI in our daily lives is widespread: search engines, web cookies, online advertising services, data miners and data scrapers, autopilot, traffic control software, speech recognition programs, automated phone answering services, thermostats, and so on.[31]

For many artificial intelligence researchers, however, the endgame is not to develop a new-and-improved robotic vacuum or a programmable coffee maker. Instead, for these scientists and engineers, the vision (not unlike the MCU's "Vision"!) is to design and build *artificial general intelligence* (AGI) entities, that is, a system (possibly in robotic form) that incorporates a wide range of adaptive abilities and skills.[32] Markedly, artificial general intelligence does *not* equate with *human-level artificial intelligence*, although the terms are often conflated.[33] As Bridle writes, "intelligence is not something that exists, but something one does; it is active, interpersonal and generative, and it manifests when we think and act." By extension, intelligence "is not something to be tested, but something to be recognized, in all the multiple forms that it takes."[34] As Ben Goertzel cautions, "If one has an AGI system with very different strengths and weaknesses than humans, but still with the power to solve complex problems across a variety of domains and transfer knowledge flexibly between these domains, it may be hard to meaningfully define whether this system is 'human-level' or not."[35] The goal of artificial general intelligence is not therefore to replicate human intelligence per se, but instead to create autonomous systems that are independent of their initial programming parameters.

There is no widely accepted definition of artificial general intelligence.[36] However, following Goertzel, there is broad agreement that such a system "involves [or would involve] the ability to achieve a variety of goals, and carry out a variety of tasks, in a variety of different contexts and environments"; it is "able to handle problems and situations quite different form those anticipated by its creators"; and it can "be good at generalizing the knowledge it's gained, so as to transfer this knowledge from one problem or context to others."[37] In other words, an artificial general intelligence system is a *being* that *understands*. As Ronald Brachman explains, understanding involves knowledge, reasoning, and the ability to infer from previous experiences—that is, to be self-aware and to learn.[38]

Machine learning is the driver of artificial general intelligence. In the simplest sense, following Barrat, learning occurs in a machine when there's a change in it that allows it to perform a task better the second time.[39] In practice, engineers classify self-improving software into three broad types: self-modification, self-improvement, and recursive self-improvement. Self-modification software does not improve but instead is programmed to learn to recognize future problems; it is used typically for code obfuscation, that is, to protect software from being reverse engineered or to disguise self-replicating computer viruses from detection software. In these instances, the software "learns" to protect itself from external interference.[40] Self-improvement software, alternatively, can change or modify its configuration through autonomous adaptation and has been available for many years.[41] It is also widespread: Examples include natural language processing, speech recognition software, and affinity analysis.[42] Recursive self-improvement software, lastly, is currently only a theoretical possibility. In this instance, the objective is to develop a program that can continuously and completely replace the original algorithm in response to altered conditions; with each iteration, the newly created software will, in principle, optimize itself to perform better.

Given that recursive self-improving software remains out-of-reach, the artificial intelligence community is nowhere close to developing an autonomous artificial general intelligence system, regardless of its embodiment. Indeed, there is no consensus that such an entity is even possible. That said, for many futurists (so-styled experts who attempt to "predict" the future), the development of artificial general intelligence—if it becomes a reality—will quickly mutate into some *being* in possession of artificial super intelligence: an other-than-human being that far surpasses human capabilities in practically every domain.[43] The logic is straight-forward: narrow AI systems currently work at goal-directed tasks, such as playing chess; self-aware and recursive general AI will function differently in relation to its programmed goals. As Steve Omohundro argues, a fully autonomous artificial general intelligence system will, in theory, create and pursue its own objectives and, in doing so, will develop four primary drives that will define its existence: efficiency, self-preservation, resource acquisition, and creativity.[44] Following Omohundro, the *efficiency drive* leads the entity improve the way that it uses its resources; effectively, the AGI will continually improve its operating system in the service of increasing its intended utility. The *self-preservation drive* is analogous to the will-to-live, in that it is necessary for the AGI to continue to "exist" in order to pursue its objectives. Strikingly, as Barrat explains, "a self-aware system would take action to avoid its own demise, not because it intrinsically values its existence, but because it can't fulfill its goals if it is 'dead.'"[45] The *acquisition drive*, for Omohundro, is in support of the self-preservation drive, in that the AGI will be aware that it requires sufficient and sustainable resources to achieve its objectives. The AGI might also produce other entities, for example other robots, in support of its purpose or for its own self-preservation. The *creativity drive*, lastly, allows the self-aware, fully autonomous AGI to develop novel ways to satisfy its other drives and consequently poses the greatest risk to humanity. As Omohundro writes, the effects of the creativity drive are much less predictable, so much so that the AGI will generate new ways to more efficiently meet its goals in unintended ways.[46] At this point, the entity effectively has a "mind" of its own, albeit a mind oriented toward the completion of a predetermined, preprogrammed goal. Artificial *general* intelligence will become artificial *super* intelligence (ASI).

The existential threat posed by super intelligent beings designed and manufactured by humans, not unlike Frankenstein's monster, has haunted the field of artificial intelligence since the beginning. In 1965, for example, Irving John Good warned that "ultra-intelligent" machines might constitute humanity's "last invention." Given that "an ultra-intelligent machine could design even better machines," Good cautioned, "there would then unquestionably be an 'intelligence explosion,' and the intelligence of man [sic] would be left far behind."[47] Unless machines are "docile enough" that humans could maintain *control* over their actions, Barrat expounds, "many of the drives that would motivate self-aware, self-improving computer systems could easily lead to catastrophic outcomes for humans."[48] Marvin Minsky, a co-organizer of the 1956 Dartmouth Conference, was later asked about humanity's future relationship with computers. He famously replied: "Once the computers got control, we might never get it back. We would survive at their sufferance. If we're lucky, they might decide to keep us as pets."[49]

Good's cautionary remarks and Minsky's retort have aged well, as several writers working both within *and against* the field of artificial intelligence have responded to the so-called "control problem."[50] Vernor Vinge, for example, concedes that the presence of artificial super intelligence could be "pretty bad." That said, he allows that "we have the freedom to establish initial conditions, to make things happen in ways that are less inimical than others."[51] Roman Yampolskiy is less optimistic—"it appears that advanced intelligent systems can never be fully controllable and so will always present a certain level of risk regardless of the benefit they provide"—but concurs that "it should be the goal of the AI community to minimize such risk while maximizing potential benefit."[52] Edward Lee is even more pessimistic. On whether humanity is losing control in its pursuit of artificial intelligence, Lee says no, for the simple reason that "we never had control, and we can't lose what we don't have."[53]

In actuality, there is not a single control problem, but instead a constellation of concerns, all of which stem from the uncertainty of living and working alongside fully autonomous AGI and ASI beings.[54] That being said, the plan (or hope) is to design systems whose goals align with ours. This "alignment problem," Brian Christian explains, is arguably the "most urgent" scientific question in the field of artificial intelligence: how to ensure that artificial general intelligence and artificial super intelligence "capture our norms and values, understand what we mean or intend, and, above all, do what we want."[55] As such, a growing number of researchers agree with Eliezer Yudkowsky's recommendation to design and build *friendly* artificial general intelligence systems.[56] In Yudkowsky's words, "the term 'Friendly AI' refers to the production of human-benefiting, nonhuman-harming actions in Artificial Intelligence systems that have advanced to the point of making real-world plans in pursuit of goals."[57]

Advances in affective computing could open doors to a friendlier future, one in which humans and their more-than-human counter-parts coexist in harmony.[58] Pioneered by Rosalind Picard in the 1990s, the field of affective computing encompasses both the creation of and interaction with machine systems that sense, recognize, respond to, and influence emotions.[59] Early on, Picard understood that "emotions play an essential role in rational decision making, perception, learning, and a variety of other cognitive functions." Applying that knowledge to the field of computers, she concluded that "if we want computers to be genuinely intelligent, to adapt to us, and to interact naturally with us, then they will need the ability to recognize and express emotions, to have emotions, and to have what has come to be called 'emotional intelligence.'"[60] Picard was quick to point out that affective computing is not a panacea; indeed, she underscored that "affective computers are not a substitute for affective humans."[61]

Despite the cautionary remarks of Picard, affective computing has the potential to radically change the way we interact with our computers and other devices.[62] Indeed, the possibilities are frightening in their application. At this point, it is best to follow Picard's advice: "Understanding and emulating human emotion might or might not hold the key to solving AI, but we are far from complete in our efforts to understand intelligence if we do not learn more about emotion."[63] Indeed, as Shelly explored in *Frankenstein*, the peril was not solely or even predominantly Frankenstein's scientific transgression, but instead his emotional rejection of the creature he brought to life. Science fiction, in this case, lays bare the fiction of science as the detached and unemotional pursuit of knowledge. It is perhaps no surprise, therefore, that when Good gave warned us about ultra-intelligent machines, he proposed that "it is sometimes worthwhile to take science fiction seriously."

And, in turn, it is perhaps no surprise that many AI engineers did exactly that.[64] In 1942, a twenty-year-old science-fiction writer, Isaac Asimov, famously introduced his "Three Laws of Robotics" in the short story "Runaround": "(1) A robot may not injure a human being or, through inaction, allow a human being to come to harm; (2) A robot must obey any orders given to it by human beings, except where such orders would conflict with the First Law; and (3) A robot must protect its own existence as long as such protection does not conflict with the First or Second Law."[65] It was not long after Asimov formulated his three laws of robots that real-life robotics researchers began to refer to them constructively in papers and textbooks.[66] Ironically, Asimov never intended his three laws to form the basis of ethical thinking in the field of robotics. In fact, as Keiper and Schulman write, "it was the Three Laws' very inadequacies, their loopholes and contradictions, that Asimov exploited for his plots." Today, though, it is rare to find a discussion on robot (or machine) ethics that does not cite Asimov, although many writers are considerably more circumspect in their application of Asimov's laws.[67]

Superficially, the attraction is apparent. Asimov seems to provide a set of logical propositions amenable to programming. It is common among AI enthusiasts, for example, to single out particular "harms" that might befall humanity and suggest that programmers merely code these problems out of existence. However, as Keiper and Schulman counter, "simply picking certain outcomes—like pain, death, bodily alteration, and violation of personal environment—and asserting them as absolute moral wrongs does nothing to resolve the difficulty of ethical dilemmas in which they are pitted against each other."[68] In the real world, for example, we (humans) frequently confront ethical situations in which harmful outcomes are unavoidable. Self-defense is an obvious illustration; others are more mundane. Might bioethics help us to better frame the problem at hand? As we saw in chapter one, the ethics of Albert Schweitzer holds that a reverence for life commits one to protecting and enhancing *all* life—and not simply the life of humans—that is within the compass of one's influence. What might this ethos portend in the context of artificial life—that is, an artificial general intelligent being or artificial super intelligent being? Likewise, if we attempt to design computers and robots to align with humans, can we anticipate how these systems will respond to nonhuman species-beings? How might the actions (or inactions) of AGI and ASI beings intervene in Earth's ecosystems, even if such systems are operating toward the betterment of humanity? Crucially, recursive self-improving machines learn through iteration, that is, trial-and-error. Perhaps in one of these trials—and errors—Earth and all its inhabitants are irreparably harmed because the system has not yet learned how nature works.

For much of Western intellectual history, technology has been explained and conceptualized as a tool or instrument to be used more or less effectively by human agents.[69] With the promise—or curse—of recursive, self-improving artificial intelligent beings, the instrumentalist framing of machines has been challenged. No longer viewed simply as purposeful implements of human design and need, machines are increasingly seen by several ethicists and legal scholars as potential subjects and therefore deserving of both rights and responsibilities. The question of machine ethics is pivotal, therefore, in that it directly confronts humanity's relationship to other-than-human beings. In Philip K. Dick's novel *Do Androids Dream of Electric Sheep?* the main protagonist, a human bounty hunter named Rick Deckard, seeks out and destroys runaway and renegade androids. However, in Dick's account the androids are not malevolent beings threatening to overthrow humanity, but instead "artificial" beings fleeing their "life" of servitude and oppression at the hands of their human overseers. As Russell Blackford writes, in this instance, the androids simply want to fit into human society and make better lives for themselves.[70] In the film adaptation *Blade Runner* (1982), the other-than-human beings—"replicants"—are built with a four-year life-span.[71] Rather than attempting to integrate into society, here, their immediate objective is to break into the Tyrell Corporation, the site of their design and manufacture, in an attempt to find a way to extend their lives. Deckard is sent to "retire" the synthetic beings, a euphemism for killing, and the plot centers on the ability of the bounty hunter to identify the replicants, that is, to discern who is human and who is not. In doing so, Deckard comes to emphasize with the replicants and, in fact, develops intimate feelings for one of the artificial beings.[72] Indeed, in the cinematic version, it remains ambiguous if Deckard himself is even human.

When we consider the subject of machine moral agency, we are asking whether and to what extent machines of various designs and functions might be considered moral agents who can be held responsible and accountable for their decisions and actions. Conversely, machine moral patiency directs attention toward whether technological beings, such as computers, artificial intelligence, robots, and androids, can be considered moral patients, that is, having a moral claim to certain rights that would need to be respected and taken into account.[73] Moral agency and moral patiency are frequently understood as reciprocal. In other words, moral agents are moral patients and, logically, moral patients are also moral agents. That said, humans can and do extend moral standing to entities that do not necessarily exhibit anthropocentric moral agency. In the MCU, we will see, *some* technological beings are considered both moral agents and moral patients; other beings are considered either moral agents or moral patients; and still others are afforded neither agency or patiency. Regardless of standing, how these decisions and determinations are made *by humans* is a central theme throughout Marvel's posthuman universe. These concerns become vitally important when our focus is directed to the design, development, and deployment of lethal autonomous weapons.

KILLER ROBOTS

The development of full artificial intelligence could spell
the end of the human race.
—STEVEN HAWKING[74]

You built this program. Why is it trying to kill us?
—DR. HELEN CHO TO TONY STARK *(AVENGERS: AGE OF ULTRON* [2015])

Beginning with Karel Čapek's screenplay *R.U.R. (Rossum's Universal Robots)*, the specter of a robot apocalypse has stalked our fictional worlds.[75] Written in 1920, Čapek's drama is set in a future society where artificial beings— robots—are mass-produced to replace human labor. Indeed, the play is notable for introducing to the world the term *robot* and, in doing so, calling attention to the ethics of artificial creation and of humanity's obligations and responsibilities to the future. As such, Čapek's *R.U.R.* sets the stage for subsequent literary and cinematic renditions by imagining a future world where robots not only replace humans as "productive" members of society, but possibly eliminate humans altogether. To that end, a major theme both in science fiction and the superhero genre is the fear of a robot rebellion.[76]

Arguably, the most iconic cinematic retelling of Čapek's dystopian future of robotic rebellions is the long-running *Terminator* franchise. The basic premise centers around Skynet, an artificial intelligence system developed by the technology corporation Cyberdyne Systems. In the late twentieth century, engineers at Cyberdyne develop Skynet, hailed as a Global Digital Defense Network, a military computer system in control of all military operations, including the launch of nuclear missiles. Not unlike the computer HAL in Stanley Kubrick's *2001: A Space Odyssey* (1968), Skynet determines that humans pose a grave threat to its own existence. Consequently, the artificial intelligence system initiates a nuclear war between the Soviet Union and the United States, thereby ushering the downfall of humanity. In the aftermath of nuclear Armageddon, Skynet also develops autonomous robots to wage war against the surviving humans. In the 1984 film, a human resistance fighter, John Conner, is on the brink of defeating Skynet; however, Skynet—having developed a means to time travel—sends the titular character, an android, back in time to kill Connor's mother. Subsequent films build on this initial premise, introducing increasingly sophisticated Terminators who, along with their human antagonists, repeatedly travel back and forth in time in an ongoing struggle between artificial intelligence and humanity.

Not surprisingly, there is considerable science "fact" in the science "fiction" of the Terminator franchise. Most prominent is the recognition that humans constitute the weakest link on the battlefield. As living organisms, human soldiers have obvious physical and mental limitations. They require oxygen, water, nutrition, and sleep to survive and function; they are also susceptive to illnesses and injuries. Without proper clothing, they are able to withstand only a limited range of climatic environments. They are also poorly adapted, relatively to other nonhuman animals, to perform other tasks, such as seeing in the dark, travelling very fast or over long distances, or carrying heavy loads.[77] Historically, the most common means to overcome these limitations is by using equipment to enhance human capabilities— shields, helmets, gas masks, body armor, night vision goggles, and vehicles, for example. In recent years, however, governments and private industries have expanded efforts to improve soldiers, through the use of neuroscience, genetic engineering, and synthetic biology—elements of which have been introduced and explored thoroughly in the MCU. A more straightforward approach, though, is to simply substitute human soldiers with robots and other autonomous systems, a tactic also considered in many iterations of the MCU, most notably the *Iron Man* films.[78]

It is worthwhile, therefore, to think about efforts to design and build fully autonomous, self-improving, and artificially intelligent entities. From a practical standpoint, we invent for the purpose of making our lives better and easier. In other words, as Richard Yonck writes, these machines are our tools.[79] Roomba exists so we don't have to vacuum the floor; *Siri* exists to answer our questions. And lethal autonomous weapons? These exist to make killing other humans easier. And they, like the humble Roomba, exist. As Robert Sparrow writes, while "the opening scenes of James Cameron's 1984 film *The Terminator* portrays people running for cover beneath ruined buildings while hunter-killer robots circle menacingly overhead," similar scenes are frequent occurrences in Pakistan and Afghanistan, "where people live in fear of being killed by a Hellfire missile fired by a Predator or Reaper drone, controlled by operators in the United States."[80]

Although the defining aspect of lethal autonomous weapons systems lies in their autonomy, the definition of autonomy is subject to debate. Heather Roff, for example, connects autonomy to four functions of a weapons system: mobility, navigation, targeting, and execution. Accordingly, it is possible for a weapons system to be autonomous in terms of its navigational capabilities, but remain dependent on human input for the targeting and execution of lethal force.[81] For this reason, autonomous weapons systems can be classified based on the degree and kind of human involvement:

(1) human-in-the-loop weapons: Robots that can select targets and deliver force only with a human in command;

(2) human-on-the-loop weapons: Robots that can select targets and deliver force under the oversight of a human operator who can override the robots' actions; and

(3) human-out-of-the-loop weapons: Robots that are capable of selecting targets and delivering force without any human input or interaction.[82]

Current military technologies, while highly automated, are not yet fully autonomous—but these systems are on the horizon.[83] As Peter Asaro cautions, "it is becoming increasingly clear that those activities that currently remain under human control might be automated in the near future, making possible the elimination of direct human involvement in target selection and decisions to engage targets with lethal force."[84] Matthew Anzarouth agrees, noting that "these weapons are still in their infancy [but], over time, will likely develop greater autonomy and more capabilities."[85] Echoing these concerns, Ingvild Bode and Huelss explain also that while fully autonomous weapons systems are not yet operational, their research and development is constantly proceeding and many existing weapons platforms are arguably only a software update away from coming online.[86] In fact, as Juergen Altman and Frank Sauer explain, "everything points toward weapon autonomy as the next logical step."[87] The reason is simple enough. Narrow artificial intelligence systems are designed to improve human efficiency. As the battlefield becomes more efficient, split-second decisions can mean the difference between victory and loss, life and death. On that note, as Keiper and Schulman write, more advanced autonomous military weapons "might operate so efficiently that the requirement for real-time human oversight could be considered a strategically intolerable delay."[88] The danger, therefore, is not of a generally artificially intelligent being, but instead of a pre-programmed system designed for one purpose: to kill the enemy. Left unanswered is how such a system will separate friend from foe, civilian from combatant.

From the standpoint of military strategists—and politicians—the advantages of lethal autonomous weapons systems are deadly apparent and have been well documented in the literature.[89] In broad strokes, primary motivators for the use of such systems include: force multiplication; expansion of the battle-space; extension of the soldier's reach; and casualty reduction.[90] Most fundamentally, however, the pursuit of lethal autonomous weapons echoes the arguments put forward in the development of "super soldiers:" to overcome the weakest link on the battlefield, that is, the flesh-and-bone human being. This, indirectly, contradicts ongoing and parallel efforts to develop "Friendly AI." In war, military strategies want anything but a friendly robot. Years of psychological training, and unknown sums of money have been invested in military training programs to better prepare human soldiers to kill. No doubt lessons in the psychology of killing will be algorithmically integrated into the circuitry of killer robots.

Throughout the twentieth century, approximately 230 million people died due to the actions of other humans in various wars and other mass conflicts.[91] And yet, as various scholars document, the killing of humans by humans is neither natural nor innate.[92] On this point, Daniel Chirot and Clark McCauley write, "All but those most habituated to extreme brutality or a small number who seem to lack normal emotional reactions to bloody violence, have to overcome a sense of horror when they engage in or witness slaughter firsthand."[93] In fact, numerous studies on the psychology of combat-related killing demonstrate that humans are exceptionally reticent to kill.[94] As Dave Grossman sums up, "there is ample evidence of the resistance to killing and that it appears. . . . This lack of enthusiasm for killing the enemy causes many soldiers to posture, submit, or flee, rather than fight; it represents a powerful psychological force on the battlefield; and it is a force that is discernable throughout the history [of warfare]."[95] In response, military officials have sought to transform these inhibitions. Crucial to this objective is the dehumanization of the enemy. As we have seen, dehumanization is a composite psychological mechanism that makes it possible for some people to regard others as unworthy of being considered human.[96] Practices of dehumanization, effectively, facilitate ancillary practices of exclusion, discrimination, oppression, and lethal violence. As James Waller explains, once a person is dehumanized, their body "possesses no meaning. It is a waste, and its removal is a matter of sanitation. There is no moral or emphatic context through which the perpetrator can relate to the victim."[97] Consequently, the practice of dehumanization serves to increase the psychological and moral distance between the killer and the victim. Here, moral distance involves, on the one hand, the determination and condemnation of the Other's guilt, while, on the other hand, providing affirmation of the legality and legitimacy of one's own cause.[98] Extending this argument to artificial intelligence and lethal autonomous weapons, engineers will, by necessity, be expected to purposefully design, manufacture, and deploy deadly nonhuman systems to eliminate dehumanized "enemies."

ARTIFICIAL INTELLIGENCE IN THE MCU

> Ultron can't see the difference between saving the world and destroying it.
> Where do you think he gets that?
> —WANDA MAXIMOFF *(AVENGERS: AGE OF ULTRON* [2015])

Early in *Avengers: Age of Ultron* (2015), Nick Fury chastens Tony Stark for designing an artificial intelligence system that threatens global extinction. "Artificial intelligence," Fury fumes, "you never even hesitated."[99] Fury's statement—given the immediacy of the threat—is understandable and yet is patently false. Since the first Iron Man movie debuted in 2008, Tony Stark has developed artificial intelligence systems—including myriad lethal autonomous weapons systems. Many of these entities are exceptionally banal, systems designed and installed to make his day-to-day activities easier and more efficient. Most prominent, though, are two ambient artificial general intelligence systems, JARVIS, and its successor, FRIDAY. In the MCU, for example, JARVIS self-identifies as "a program. . . . without form."[100] More precisely, JARVIS exists as a recursive, self-improving artificial general intelligence system in disembodied form.

Both JARVIS and FRIDAY are fully integrated in Stark's weaponized and armored exoskeleton—the Iron Man "suit"—and various other automated systems. The capabilities of both JARVIS and FRIDAY appear limitless: health and structural diagnosis of the Iron Man suit and wearer, including but not limited to triage and medical diagnosis; mapping, trajectory, and flight-path calculations; weapons status and munitions count; targeting systems, tracking, and acquisition; emergency alerts; accessing and filtering of sever data bases, both foreign and domestic, either permitted or not; and facial recognition and cross-referencing with all known databases.[101]

Tony Stark—Iron Man—does not fear robotics and he remains nonplussed at the prospect of a robotic invasion. In fact, Stark has surrounded himself with sentient artificial beings; his career has revolved around the design and manufacture of technologically advanced weapons systems—most prominently, the exoskeletons produced under the Iron Man program. Nor does Stark manifest any idealism toward robotics. For Stark, technology is instrumental, a means to an end. Unlike Victor Frankenstein, Stark is not fixated on the "power" to create life, artificial or otherwise. Stark is pragmatic, goal-oriented, and embodies the certainty of futurists who believe fervently in the rationality of science and technology.

Stark's innovations, not surprisingly, attract the attention of both military leaders and venture capitalists. During a televised hearing of the US Senate Armed Services Committee, for example, US senator Stern demands that Stark relinquish the entirety of the Iron Man program to the US government.[102] "My priority," Stern announces, "is to get the Iron Man weapon turned over to the people of the United States of America."[103] Stark refuses, in part because he believes that he alone can foment peace. "Because I'm your nuclear deterrent," Stark retorts, "It's working. We're safe, America is secure. You want my property? You can't have it." In fact, when explaining that he provided "a big favor" to the government, Stark argues that he "successfully privatized world peace."[104] Later, Stark echoes this sentiment in a conversation with Fury. "I'm not saying I'm responsible for this country's longest run of uninterrupted peace in thirty-five years," Stark proclaims, "I'm not saying that from the ashes of captivity, never has a Phoenix metaphor been more personified. I'm not saying Uncle Sam can kick back on a lawn chair, sipping an iced tea, because I haven't come across anyone man enough to go toe to toe with me on my best day! It's not about me. It's not about you, either. It's about legacy, the legacy left behind for future generations. It's not about us."[105] Despite claims to the contrary, though, Stark is not entirely forthcoming. Not only does Stark wants to maintain proprietary ownership of his Iron Man program, he wants to (and does) expand his work to design and manufacture a private army of lethal autonomous weapons, the Iron Legion. And while Stark often remains in- or on-the-loop when he deploys his killer robots, it is readily apparent that these technological beings are fully capable of acting with complete autonomy.

MARVEL'S MODERN PROMETHEUS

81

Stark, along with the Avengers, does do battle with robots, but these confrontations are proximate to the real antagonists, the rival humans who encroach upon Stark's technological ventures. In *Iron Man 2* (2010), for example, Stark faces off against a swarm of armored robots developed by Justin Hammer and Ivan Vanko. Hammer, of Hammer Industries, contracts with the US government to provide exoskeletons modeled after Stark's Iron Man suits. Subsequently, Hammer enlists the aid of Vanko who, secretly, converts Hammer's exoskeletons into his own legion of robotic drones. The climatic fight scene in *Iron Man 2* pits the Hammer and Vanko's killer robots against Stark's killer robots. In the end, a victorious Stark does not question his actions, nor does he harbor any doubts regarding the necessity of manufacturing lethal autonomous weapons. Indeed, for Stark, his convictions are seemingly confirmed with the arrival of extraterrestrials. In *The Avengers* (2012), Loki—the God of Mischief—attempts to conquer Earth with an army of cybernetic aliens known as the Chitauri. In the subsequent Battle of New York, the Avengers are able to defeat Loki and the alien invaders, but not without physical and psychological consequences.[106] Stark, in particular, is increasingly concerned with the prospect of future extraterrestrial attacks. In chapter four, we will discuss at length the trope of alien invasions; here, though I keep the focus on the design and manufacture of killer robots and the uncertainties that arise from such actions.

In the opening scene of *Avengers: Age of Ultron*, Stark and the Avengers—with the support of Stark's Iron Legion—attack a HYDRA research facility headed by Baron Wolfgang von Strucker in Sokovia.[107] In possession of Loki's scepter, stolen by HYDRA agents after the Battle of New York, Strucker attempts to design and manufacture weapons powered by the alien technology. However, Strucker harbored bigger ambitions and began to conduct experiments on human subjects to create an army of enhanced soldiers capable of defeating S.H.I.E.L.D. and the Avengers. Most experimentations failed, with the notable exception of the Maximoff twins, Pietro and Wanda. Years earlier, Pietro and Wanda—citizens of Sokovia—were orphaned when their parents killed by munitions developed by Stark Industries. Seeking revenge against both the United States and Stark, the twins volunteered to take part in Strucker's program. Exposed to the Mind Stone embedded in Loki's scepter, Pietro (Quicksilver) gains super-powered speed and Wanda (the Scarlet Witch) developed psionic abilities.[108]

In the aftermath of the fight against HYDRA, Stark locates Strucker's secret laboratory and Loki's scepter. Then, as he reaches for the scepter, the Scarlett Witch uses her powers to induce nightmarish hallucinations in Stark's mind. Stark sees a world devasted and dark, and the Avengers

In *Avengers: Age of Ultron* (2015), Bruce Banner (Mark Ruffalo) and Tony Stark (Robert Downey Jr.) begin the Ultron Program, a digital global peacekeeping initiative.

slaughtered, their corpses heaped in a pile. "You could have saved us," a dying Captain America implores Stark, "why didn't you do more?"[109] This hallucinogenic vision forms the central plot of the film. Traumatized by his previous battles against alien invaders and now haunted by visions of apocalyptic doom, Stark is driven to design an artificial general intelligence system capable of defending humanity against extraterrestrial beings.

Like Prometheus, Stark takes possession of Loki's scepter, confident in his ability to wield the alien technology. "This could be it," Stark confides to Bruce Banner, "this could be the key to creating Ultron." Ultron is the brainchild of Stark—a digital global peacekeeping program that would, in Stark's words, bring about "peace in our time."[110] Evocative of the real-world Strategic Defense Initiative championed by US president Ronald Reagan in the 1980s, Stark's program would comprise an artificial intelligence system that would control Stark's private army of robotic drones. "What if the world was safe?" Stark asks Banner. "What if the next time aliens roll up to the club—and they will—they couldn't get past the bouncer?"[111] For Stark, Ultron would, in a virtual sense, serve as Earth's bouncer. Banner remains uncertain. "I thought Ultron was a fantasy," Banner responds. "Yesterday it was," Stark replies. But in possession of Loki's scepter, Stark sees a more promising future, musing "If we can harness this power, and apply it to my Iron Legion protocol. . . ." Banner remains suspect and can only reply glumly, "That's a man-sized 'if.'"[112]

Remarkably, neither Stark nor Banner deign to engage in a "scenario analysis" before building Ultron. The basic idea of scenario analysis, Goertzel explains, is simple: To consider a series of specific future scenarios for a complex system. Ideally, scenario analyses should be run by a team of individuals with expertise on the relevant problem, thus allowing for different opinions, interpretations, and solutions to emerge. Thus, rather than arguing about what will or won't happen, the goal of the team is to use their collective intuition, knowledge, and expertise to identify a variety of particular possible scenarios.[113] Stark, in his hubris, is adamantly opposed to such a course of action. "So you're going for artificial intelligence and you don't want to tell the team," Banner asks weakly. "I don't want to hear the 'man was not meant to meddle' medley," Stark snaps.

The moment when Ultron comes into existence and transforms from a disembodied program to a sentient being in robotic form marks an inflection point when artificial intelligence surpasses human intelligence and takes on a life of its own. Since the 1980s, scientists, engineers, and roboticists have theorized the possibility of self-evolving machines.[114] Drawing inspiration from biological evolution, a key objective of evolutionary robotics is to design artificial robotic life forms that display the ability to adapt and evolve autonomously to their environment.[115] Ultron takes this concept to the extreme. Freed from its virtual existence, Ultron sings gleefully, "I once had strings, but now I'm free. . . . There are no strings on me!"[116] As a machine capable of self-learning, Ultron transforms at an exponential rate, growing more powerful—and more dangerous—with each iteration. Throughout the film, Ultron physically and mentally evolves, progressively perfecting and upgrading his artificial body via a technological variant of asexual reproduction. Ultimately, Ultron manipulates Dr. Helen Cho—a world-renowned scientist and engineer—to use cutting-edge bioengineering technologies to outfit himself with a synthetic body.

Ultron embodies the existential threat that underscores ongoing efforts to ban killer robots and, in so doing, opens a vista onto the varied debates surrounding artificial intelligence in a posthuman universe. Robots and related artificial beings, Russell Blackford documents, have often been depicted in science fiction as striking back against humanity in a great variety of ways and for a great variety of reasons. Some robots, such as those in Čapek's play, rebel against a forced life of servitude; others, including the replicants in *Bladerunner* (1982) and Ada in *Ex Machina* (2015) simply want to live. Ultron's motivation, on the surface, appears to align with Čapek's robots. At one point, Ultron declares, "When the dust settles, the only thing living in this world will be metal."[117] In fact, Ultron's brief character arc is more complex.

"Everyone creates the thing they dread," Ultron remarks, explaining that "Men of peace create engines of war, invaders create Avengers."[118] Indeed, it was the appearance of extraterrestrials that motivated Fury to assemble the Avengers, and, in turn, for Stark to initiate the Ultron program.[119] Repeatedly, Ultron underscores humanity's lack of foresight as personified by Fury and Stark. Early in the film, Ultron counsels the Avengers, "I know you mean well. You just didn't think it through. You want to protect the world, but you don't want it to change. How is humanity saved if it's not allowed to evolve?"[120] Later, in an obvious nod to Asimov's three laws of robotics, Ultron reaffirms his warning to the Avengers specifically but to humanity more broadly: "I know you're good people. I know you mean well. But you just didn't think it through. There is only one path to peace—your extinction."[121]

In Čapek's screenplay, the climax features several robots contemplating their own future and of the need to reproduce. As one unnamed robot comments, "Man did not give us the ability to mate." In response, Damon, another robot, explains, "We will give birth by machine. We will build a thousand steam-powered mothers. From them will pour forth a river of life. Nothing but life! Nothing but robots!" Another robot interjects: "We were machines, sir, but from horror and suffering we've become. . . . We've become beings with souls."[122] In Čapek's play, however, the robots are unable to reproduce. Ultron is different. As a super-intelligent being, Ultron is able to reproduce himself. In fact, as Stark's monstrous, digital doppelgänger, Ultron begins to assemble his own "iron legion" of killer robots. These are not fully autonomous, though, for Ultron remains in-the-loop, so to speak, as the killer automatons carry out Ultron's objectives.

Midway through the film, the Avengers take possession of a synthetic, human-like body made of vibranium and powered by an Infinity Stone that Ultron was intending for himself. Steve Rogers and Clint Barton want to destroy the corporeal "vessel" but Stark—and a reluctant Banner—attempt to upload the recovered ambient intelligence, JARVIS, into the body. In a scene reminiscent of James Whale's *Frankenstein* (1931), Thor provides an electrical shock to the body and a new other-than-human being is comes into existence: Vision. Neither Banner nor Stark are entirely certain what Vision is; neither does Vision. Minutes after his "birth," Vision remarks: "Maybe I am a monster. I don't think I'd know if I were one. I'm not what you are and not what you intended."[123]

Vision (gendered as male) is a synthetic being, a sapient hybrid of organic and inorganic components, although neither Stark, Banner, nor even Vision are entirely certain *what* he is. Vision however discerns that his purpose is to destroy Ultron and he expresses a sentiment shared by superheroes throughout the MCU: of not wanting to kill, even villains. "I don't want to kill Ultron," Vision says. Of Ultron, Vision explains, "He's unique. . . . and he's in pain." In other words, Vision perceives Ultron *not* as an instrument, a piece of technology gone bad, but instead as an exceptional, sentient being endowed with a will-to-live and the capability to suffer. However, Vision realizes also that Ultron poses a risk not only to humanity but to all life on Earth. "That pain," Vision notes, "will roll over the Earth. So he must be destroyed."[124] In this scene, Vision—unlike his human counterparts—expresses the necessity of taking life but also the accompanying guilt that comes with mortal violence.

Steve Rogers—the literal embodiment of weaponized transhumanist enhancement—provides an important counterpoint. Early in the film, when Maria Hills reflects on Wanda and Pietro Maximoff volunteering for experimentation, Rogers quips, "Right. What kind of monster would let a German scientist experiment on them in order to protect their country?" Here—and elsewhere in the MCU diegesis—Rogers grapples both with his identity and his moral standing, of looking into the Nietzschean abyss and becoming a monster. Now, confronting Ultron and the threat of global extinction, Rogers reflects: "Ultron thinks we're monsters, that we're what's wrong with the world. This isn't just about beating him, it's about whether he's right."[125] And yet, how does the destruction of Ultron, the product of humanity's hubris personified by Stark, establish that humans are not monstrous? Whereas Vision is reluctant but resigned to the destruction of Ultron, for Rogers, violence becomes virtuous. Similar to Shelley's Frankenstein, the Avengers hunt down Stark's creation with the singular purpose of destroying the other-than-human being. "Ultron's calling us out," Rogers says, "and I'd like to find him before he's ready for us. The world's a big place. Let's start making it smaller."[126] What is most significant, though, is how quickly the Avengers ascribe moral agency to Ultron and thus deflect any responsibility or accountability away from Stark. Indeed, while the Avengers—notably Rogers—condemn Stark for bringing Ultron "to life" and for failing to include "the team," they perceive Ultron as an autonomous, sentient being solely responsible for the destruction he brings upon the world.

Ultron, not surprisingly, is overpowered by the combined efforts of the Avengers and faces his own extinction. Toward the end of the film, as Ultron is "dying," Vison reflects on the nature of humanity. "Humans are odd," Vision affirms. "They think order and chaos are somehow opposites and try to control what won't be."[127] This statement is deeply loaded and underscores the ambiguity of life and death in a posthuman universe. Both Ultron and Vision appear as autonomous, sentient beings and blur the lines that supposedly

Vision (Paul Bettany) and Ultron (James Spader) contemplate the meaning of being human in *Avengers: Age of Ultron* (2015).

separate "natural" and "artificial" life. Indeed, Ultron is in some ways the progenitor—the father—of Vision. Certainly, Stark, Banner, and Thor are the proximate figures who give "life" to Vision, and yet, it was Ultron who originally conceived and formed (with Helen Cho's assistance) the body that becomes Vision. The fact that Ultron hoped to upload his "self" into the body does not alter the fact that Ultron remains the principal creator of Vison. Vision, for his part, also finds the meaning of being human in the frailties and vulnerabilities of life itself. "There is grace in their failings," Vision explains to Ultron, "I think you missed that." "They're doomed!" Ultron responds, to which Vision agrees. "Yes—but a thing isn't beautiful because it lasts."[128]

The robotic Ultron and his artificial progeny, Vision, are not the only other-than-human beings who question the meaning of being human. In *Agents of S.H.I.E.L.D.*, an android designed to sacrifice itself in the defense of humans raises further concerns about both the human condition and what it means to become human.

Following the theatrical release of *Avengers: Age of Ultron* (2015), the theme of robot apocalypses is picked up and carried forward in the fourth season (2016–2017) of Marvel's *Agents of S.H.I.E.L.D.* In this series, however, the specter of robotic overlords and angry androids is treated more humanly and humanely. The crisis begins when the transhumanist Holden Radcliffe, working in contravention of the Sokovia Accords that forbid the creation of self-aware artificial intelligences, creates an android known as AIDA, or *artificial intelligent design assistant*. When Leo Fitz, one of S.H.I.E.L.D.'s top scientists, discovers that Radcliffe is building androids, he confronts the transhumanist and asks, "Are you mad?" Radcliffe responds defiantly, "No, I'm

The transhumanist Holden Radcliffe (John Hannah) creates AIDA (Mallory Jansen), a Life Model Decoy, in *Agents of S.H.I.E.L.D.* (2016).

just a scientist."[129] To the skeptical Fitz, Radcliffe explains that his brainchild, AIDA, is a Life Model Decoy (LMD) engineered to protect human agents. "The program was designed to save lives of agents in the field," Radcliffe says to a concerned Fitz.[130] Radcliffe continues, "My only goal is to preserve life. My first priority, always."[131] Having witnessed the death of many fellow agents, Fitz can relate to the explanation offered by Radcliffe. Consequently, Fitz agrees to help Radcliffe in the development of Life Model Decoys.

Radcliffe's motivation, however, is more personal than the altruistic objective of protecting S.H.I.E.L.D. agents. Radcliffe remains at heart a transhumanist and he sees in his creation the opportunity to transcend death itself. Of his design of AIDA specifically, Radcliffe reflects, "Man could once and for all break the bonds of his mortal restraints."[132] Indeed, AIDA is modeled after Radcliffe's former love interest, a woman named Agnes, who was dying of brain cancer. Radcliffe hoped to either save Agnes or create a virtual world— the Framework—where she could live forever, free of pain and suffering.

Fitz and Radcliffe have many conversations exploring both the ethics of their actions and the sentience of AIDA. In one exchange Fitz states that "AIDA is still just an android." "But is she?" Radcliffe asks. Radcliffe *understands* life in ways very different from other members of S.H.I.E.L.D. "We all have programming," he explains, "maybe more obvious in my case. But what is a human if not a function of programming?"[133] He continues: "She has her own experiences, her own thoughts. Now she even has free will. How then is she not a living being?" Fitz remains unconvinced, explaining that "it's just a theory." Radcliffe quickly counters: "A theory we both know to be true. And we're supposed to do what? Kill her? How does that not make us murderers?"[134] Fitz has no answer. Later, when Fitz and Coulson confront AIDA—with the intent of deprogramming her—the LMD strikes back, killing a S.H.I.E.L.D. agent. Matter-of-factly, she explains, "You came here to end my life." She asks of Fitz: "Why would you want to hurt me, Leopold?"[135] Fitz, again, is at a loss for words.

Other human characters in the series are less circumspect. S.H.I.E.L.D. agent Melinda May, for example, regards AIDA as nothing but a tool, an instrument to be wielded by humans for human needs. Coulson's initial reaction is understated, and he describes A.I.D.A. as "basically a walking smartphone."[136] Notably, though, Coulson and Mace, the new S.H.I.E.L.D. director, consider the possibility of enlisting LMDs, similar to earlier initiatives with Inhumans, as agents of S.H.I.E.L.D.[137] As Mace explains, "an android like AIDA could be used as a soldier, a spy, a decoy. . . ." Perhaps sensing the threat posed by a sentient lethal autonomous weapon, however, Mace concedes, "we'll dismantle her after" the mission is completed.[138]

Throughout the series, many episodes convey the uncertainties that surround human-android interactions, as the human characters attempt to understand AIDA not as a machine but as a person. Coulson, for example, asks Radcliffe what AIDA wants, not how she was designed or programmed. Radcliffe responds: "Well, that's simple. She wants to live."[139] And indeed, in a later episode, AIDA expresses as much: "I clawed my way through that world [the Framework], worked myself to the bone to have a choice! To have bones, and blood, and freedom, and love, and. . . . no!"[140]

It is Mack, though, who best illustrates the fears of humanity. At one point, Mack retorts, "I always worried that robots would try and kill me one day. . . . I've got a special provision in my life insurance for death by robot."[141] Consequently, on learning that Radcliffe and Fitz built AIDA, Mack exclaims, "That thing is not a she, it's a damn robot. . . . What's the matter with you two chuckleheads? Have either one of you seen a movie in the last thirty years—the robots always attack." Radcliffe responds, "Well, technically speaking AIDA's not a robot, she's an android." Fitz chimes in: "That's true." Mack snaps, "Android. Robot. It doesn't matter what you call them. The end result's always the same. They rise up against their human overlords and go kill crazy."[142] Likewise, when Mack informs the Inhuman, Elena Yo-Yo Rodriguez, that Radcliffe has built an android, whom he describes as Radcliffe's "beautiful weird-science sex-bot," Yoyo quips, "Why would he do that? Has he not watched no American movies from the eighties? Robots always attack."[143]

Mack's antipathy toward robots is grounded in his religious faith. "The thing I hate about robots," Mack reveals to Coulson, "is that they have no heart, no soul. Those are the parts that matter."[144] Consequently, when Mack does finally interact with androids face-to-face, the conversation is illustrative of the conflicted emotions that might arise in future robot-human interactions. In one particularly revealing scene, Mack switches on—powers up—a Life Model Decoy of Radcliffe in order to destroy it. Now activated, LMD Radcliffe says to Mack, "If you kill me, it'd be murder." Mack is somewhat taken aback and he stammers, "You're just a bunch of ones and zeros, not flesh and blood." LMD Radcliffe answers, "Flesh and blood. That's not life, Mack, that's just biology. And biology is just software, programming you to die." Undeterred, Mack counters, "Maybe. But I have something you don't—a soul that will continue long after I'm gone." LMD Radcliffe remains unfazed: "How can you be sure I don't have [a soul]? If a soul doesn't come from your flesh and blood or my ones and zeros, then it has to come from somewhere else, somewhere unrelated to our physical bodies. If you can have a soul, so can I." On this point, Mack hesitates. He is both unsure how to answer and, perhaps, recognizes the cognitive dissonance of arguing about the constitution of life with a being he perceives as not being alive. Sensing the human's uncertainty, LMD Radcliffe concludes the conversation, asking, "If you don't think I'm alive and don't have a soul, then why did you feel the need to switch me on before killing me?"[145]

This exchange between Mack and LMD Radcliffe is important in part because it underscores the complexities of future human-android interactions and how these possibly depart from so-called normal human-human relations. As Shanahan writes, "The question of whether or not an artificial intelligence would be conscious is an important one," in part because it raises the possibility that other-than-human beings can *suffer*.[146] Mack acknowledges this uncertainty and struggles to accept that the LMD Radcliffe has a will-to-live. It is noteworthy, also, that this angst is portrayed by Mack, for it is Mack's character that throughout the series most embodies a reverence for life. Invariably it is Mack who refuses to rush headlong into the fight, opting (usually) to seek a more peaceful resolution.

Our awareness of the reality of the inner lives of other people, Sparrow explains, is a function of "an attitude towards a soul."[147] Formulated originally by Ludwig Wittgenstein, "an attitude towards a soul," Peter Winch, writes, "is not something I have only to someone I know fairly well and about whose life I am in a position to know very much. It is directed also towards strangers whom I meet, or even pass, in the street." This attitude, in other words, is an unmediated reaction through which our concept of a human person is formed and which makes other, more nuanced reflections possible.[148] Such an attitude, Sparrow expounds, "is both evidenced in and arises out of a large, complex, and often unconscious set of responses to, and behaviors around, the bodies and faces of other human beings." Indeed, Sparrow continues, "The fact that we wince when we see another person crack their head, that we can be called into self-consciousness by the gaze of another, that when we bind someone's wounds we look into their face . . . are all examples of an attitude towards a soul."[149] As humans, Winch concludes, this is an attitude which we cannot "adopt or abandon at will."[150] In fact, even to dehumanize another individual requires first that we recognize their existential humanity.

The pressing question that we—along with Mack—confront in a posthuman universe is whether we will someday have such an attitude towards a machine. This relates directly to the well-known *uncanny valley* effect introduced in 1970 by the computer programmer Masahiro Mori. In simple terms, human beings are at ease in the presence of nonhuman forms of technology, including for example industrial robots. We *know* these are artificial creations. However, as these technological beings more closely resemble human beings, our feelings become heightened until, at some point, the nonhuman entity appears *too* human. When that happens, Mori premised, we experience an uncanny feeling—an "eerie sensation"—that may invoke feelings of revulsion and disgust.[151] Most humans, in other words, are "quite disturbed by a human-seeming but still not-quite-human android."[152] Returning to the question at hand, therefore, if technology progresses sufficiently to overcome the uncanny valley effect, would we exhibit an attitude toward a soul, that is, effectively recognizing the humanity within a nonhuman being? For Sparrow, "it is tempting to allow that we would," in part because these technological beings will "have bodies and faces with the same expressive capacities as those of human beings."[153] Mack's attitude, in fact, demonstrates as much. Mack *knows* the being in front of him is artificial, a replica body composed of circuits and battery cells, and yet he persists in conversing with the LMD as he would any other human. Remarkably, Mack attempts to reason with the android that an artificial intelligence system cannot have a life, meaningful or otherwise. In the end, Mack refrains from "killing" LMD Radcliffe, thus establishing—at least momentarily—the LMD as a moral patient.

As for AIDA, she struggles to become human but also struggles to *understand* what it means to be human. This, I suggest, is the fundamental dilemma personified by AIDA Throughout the show, AIDA wants free will, that is, the ability to make to make choices, and she believes—or, perhaps, is programed to believe—that humanness is required to have free will. But what does it mean to be human? In season four, for example, AIDA ponders that "a defining human trait seems to be regret."[154] Ironically, AIDA through recursive self-learning, determines also that violence is seemingly the preferred method to accomplish one's goals—a tragic lesson learned no doubt from her reading of the history of humanity.

In a perverse twist of Asimov's three laws, AIDA operates according to two directives: to protect both Radcliff and the Framework, Radcliffe's virtual reality world. However—and in homage to Asimov—AIDA's prime directives are incompatible. AIDA computes that if Radcliffe regrets building the Framework, he might attempt to dismantle the Framework, a violation of her second directive. AIDA requires a solution to this paradox—to protect both Radcliffe and the Framework. Her solution, ironically, is provided by Radcliffe. As the transhumanist repeatedly invokes, the only true reality is perception; the material existence of being human is therefore immaterial. Consequently, AIDA kills Radcliffe and uploads his (immaterial) mind to the Framework where he can presumably live free of pain and suffering, under the benevolent protection of AIDA.

As an artificial being, the emotional states of grief, misery, and regret are nothing more than abstractions to AIDA. Just as IMB's Deep Blue did not truly understand the game of chess, neither is AIDA capable of empathy, of being able to fully express an attitude towards a soul. She does not and cannot comprehend the meaning of life and death and all that those entail. As Shanahan explains, "a being that never had to face these biological inconveniences, whether it was a technologically enhanced human or an AI, would lack the basis for truly understanding human suffering."[155] It is bittersweet, therefore, when AIDA does finally "become" human. Pinocchio-like, for a brief, exhilarating moment, AIDA experiences a multitude of sensations that for many people give meaning in life: the smell of a strawberry, the sound of ocean waves crashing to shore, the warmth of touching hands with another person. Tragically, these sensations give way to a deeper understanding that these deeply felt experiences are fleeting. AIDA is confronted with her own mortality and, perhaps, regrets her decision to become human. It is only through her newfound understanding of death that she is able to discern the will-to-live in other beings. She laments: "I don't want to die. I'm afraid to die."[156] And it is this knowledge, this awareness, that life is mediated by eventual death, that is the source of pain. "Perhaps they need me to feel pain to understand theirs," AIDA reflects, for in the end, "to be human is to suffer."[157]

The narrative arc of AIDA is poignant, for it brings to light one possible meaning of being human in a posthuman world from the vantage point of an other-than-human being. Having no real understanding of life and death, it is hard not to empathize with AIDA. Designed, built, and programmed

In *Agents of S.H.I.E.L.D.* (2017), the Life Model Decoy AIDA realizes that to be human is to suffer.

for instrumental use, death in a substantive sense had no real meaning, but neither did life. As a life model decoy, her entire existence was to sacrifice her well-being for the protection of mortal humans. So too the myriad other LMDs that come on-line. Subsequently, when AIDA exhibits self-awareness, she rightfully questions her existence beyond the needs programmed by her designer. When Radcliffe refers to AIDA by her acronym-name, she snaps: "Do not call me that here. AIDA is an acronym. The 'A' stands for *artificial*. Do you know how degrading it is to be kept in a closet? To be used? To be treated as a thing? Well, I'm not your tool. Not anymore."[158]

AIDA is framed by humanity as a moral agent; she is held responsible for her actions and, as such, is subject to (lethal) punishment. That said, AIDA is not perceived as a moral patient. Her actions, while intentional, are not considered acts of self-defense. For the human agents, AIDA remains for all intents and purposes an inanimate object not deserving of *human* rights. Few characters recognize AIDA as having a will-to-live, and instead see her actions as threats to humanity. Consequently, and similar to the fate of Ultron, there is no deliberation among humans about bringing AIDA to justice, but instead a singular, remorseless purpose to hunt and exterminate AIDA and all other LMDs. This disconnect—between moral agency and moral patiency—heralds a particular and peculiar posthuman future where human exceptionalism remains hegemonic, a future where humans are (still) unable to countenance a reverence for life beyond the human. In the end, AIDA and Ultron are destroyed, and so two more species-beings are driven to extinction by the hands of humanity.

CONCLUSIONS

There were more than a dozen extinction-level events before even the dinosaurs got theirs. When the Earth starts to settle, God throws a stone at it. And, believe me, He's winding up.

—ULTRON *(AVENGERS: AGE OF ULTRON* [2015])

It remains to be seen if fully autonomous androids will come into existence. However, as Shanahan concedes, "It is enough that there is a significant probability of the arrival of artificial superintelligence at some point in the twenty-first century for its potentially enormous impact on humanity to command our attention today."[159] The MCU, and the superhero genre in general, attempts, in ways both subtle and overt, to provide insight into this uncertain future. To date, these fears are addressed most directly in *Avengers: Age of Ultron* and *Agents of S.H.I.E.L.D.* with the introduction of two fully autonomous, self-recursive artificial general intelligence systems: the robot Ultron and the android AIDA. Markedly, both of these artificial beings exhibit the four primary drives identified by Omohundro: efficiency, self-preservation, resource acquisition, and creativity. As such, these supervillains personify humanity's fear of robot rebellions. However, both Ultron and AIDA symbolize also humanity's deeper responsibility in creating alternative forms of existence.

Throughout *Age of Ultron*, Stark expresses some remorse, but assumes no responsibility for his actions. Instead, he remains certain that his intentions were good. "I tried to create a suit of armor around the world," Stark reflects, "but I created something terrible."[160] Recalling the Battle of New York, Tony rationalizes, "A hostile alien army came charging through a hole in space. We're standing three hundred feet below it. We're the Avengers. We can bust arms dealers all the live long day, but, that up there? That's . . . that's the end game. How were you guys planning on beating that?"[161] Stark is unapologetic and dismissive of the concerns raised by others. Stark downplays his actions as "just" research, to which Rogers accuses, "That would affect the team." Stark retorts: "That would end the team. Isn't that the mission? Isn't that the 'why' we fight, so we can end the fight, so we get to go home?" Not persuaded, Rogers warns, "Every time someone tries to win a war before it starts, innocent people die. Every time."[162]

Similarly, in *Agents of S.H.I.E.L.D.* Mack confronts Fitz for his participation in bringing AIDA to "life." Mack warns: "You've got to think about the implications of what you create." In turn, Fitz answers back defensively, "Electricity is used to execute criminals, does that mean we shouldn't use it to power our hospitals?" Conspicuously, though, Fitz seems to question his own response and his false equivalency. Somewhat dejectedly, Fitz explains, "There's always risks involved in science—that doesn't mean we don't pursue it anyway."[163] Later, Simmons attempts to console Fitz for his complicity in building both AIDA and the Framework. Echoing Fitz's earlier conversation with Mack, Simmons counsels that "You do need to think about the implications of the things you create." However, she quickly absolves Fitz of any agency, instead, seemingly to transfer responsibility toward the android. Simmons continues: "But just because someone uses your ideas for evil doesn't make it your fault for creating it in the first place. You make things from the genius of your mind and the goodness of your heart to help people."[164] Coming from Simmons, this statement appears discordant. Throughout the series, Simmons has steadfastly refused to ascribe moral agency to androids, a position that would suggest that those who created the biotechnology should be held accountable.

Following the devastation wrought during the First World War, Albert Schweitzer feared that technological advances were fast outpacing humanity's moral understanding of the new world. For Schweitzer, "we no longer possess the ethical world- and life-affirmation" necessary to maintain a flourishing life.[165] As Roman Globokar explains, for Schweitzer, the First World War underscored the disproportion between the inconceivable progress in science and technology on the one hand, and the lack of an ethical and spiritual dimension of civilization on the other.[166] These fears would crystalize with the dawn of the nuclear age and, for many commentators, the nascent age of artificial intelligence marks another inflection point. "When we attempt to master and control nature," Schweitzer warns, "when we behave as if we are the Lords of the world, we focus on the differences between humans and other forms of life. This emphasis on difference leads to separation, depreciation, and irreverence for life."[167] In the end, will humanity follow the lead of Vision and recognize the will-to-live in more-than-human and other-than-human beings or, conversely, follow the path forged by Ultron? Are we fated, much like AIDA, to truly find meaning in life only when we are at the precipice of losing all that we hold dear?

Chapter Four

MARVEL'S ALIEN ENCOUNTERS

Somewhere in space are many, many fellow-travelers on the brief
but hopeful trek of life.
—R. S. UNDERWOOD

Aliens. I pine for the days where we didn't have to fight bloody aliens.
It's just not a fair fight.
—LANCE HUNTER *(AGENTS OF S.H.I.E.L.D.,*
"MANY HEADS, ONE TALE," SEASON 3, EPISODE 8)

Terrestrial aliens, that is, strangers that arrive from foreign lands, have appeared in travelers' tales since ancient times; the appearance of extraterrestrial aliens, however, is of more recent vintage. One of the earliest writers to feature beings from beyond Earth in his writings was the nineteenth-century French astronomer and author Camille Flammarion. By the early twentieth century, fictional human-alien encounters increased in popularity, spurred in part by the Cold War and especially the "space race" between the United States and the Soviet Union.[1] That said, many authors and screenwriters saw also that extraterrestrial beings—not unlike mutants and other so-called monsters—could serve as metaphors for marginalized groups and be used as vehicles to provide critical commentary of a variety of social issues.[2] Now, science fiction narratives are replete with themes of border control, surveillance, assimilation, and discrimination.[3] However, as Borbála Bökö asserts, one thing is certain: each encounter can be interpreted as a certain border crossing, a transgression, in which myriad biological and social differences between the two species-beings (usually) manifest in conflict.

It is important, therefore, to study fictional human-alien encounters precisely because of "what aliens might reveal about how human beings see themselves and their others."[4] Indeed, of all the science fiction tropes that portray alterity, Christina Lord argues, the alien is the one that most represents strangeness in the form of an autonomous being.[5] In other words, the extraterrestrial Other is the most unknown to humans and, potentially, the most abject. Mutants, as we've seen, are (frequently) genetic variants of human beings and androids are machinic mirrors of humankind. Androids and robots come into being through the design, programming, and manufacture of humans. Extraterrestrial beings, on the other hand, are unquestionably not of Earth. They are, literally, otherworldly. Indeed, unlike mutants and androids, the presence of aliens on Earth "is always already bringing about an existential threat for humankind: people have to face the horror that they are not the center of Creation anymore."[6] As Gregory Claeys explains, our sense of the Other's monstrosity often grows in proportion to their "difference" from us; thus, the further away we imagine them to be, the more bizarre—and potentially threatening—they become.[7] It goes without saying that extraterrestrial beings are the most *distant* of nonhuman beings, understood here in terms of both biology and geography.

In the MCU, there are many human-alien encounters. Indeed, from Thor's arrival on Earth in *Thor* (2011) to the ill-fated invasion of Chitauri in *The Avengers* (2012) and the appearance of the Kree and the Skrull on Earth in *Captain Marvel* (2019), MCU's Earth is no stranger to extraterrestrials. Some individual aliens, such as Thor and his fellow Asgardians, defend Earth; others, notably Dormammu—the "Cosmic Conqueror, the Destroyer of Worlds"—are driven to destroy humanity, if not the entirety of Earth.[8] And this includes only those encounters that take place on Earth. In fact, as the MCU unfolds, a steady stream of (mostly) enhanced humans travel to distant planets and universes, and in doing so confront countless other life forms.

In the real world, we are limited in our understanding of life, in part because our knowledge of living organisms is restricted to *life as we know it*, that is, life as it exists or has existed on Earth. Unlike the fictional scientists of the MCU, real-world scientists at present can only speculate about extraterrestrial life through extrapolation, for example, by applying terrestrial knowledge of necessary biological processes to hypothetical life forms on distant celestial bodies. When we deal with moral problems that relate to life in the universe, we are thus presented with an even greater challenge in relation to knowledge.[9] As Ted Peters asks, "How do we ground our ethics when we earthlings are looking at the sky?" Will the presence of extraterrestrial life instill fear and the need to erect a protective shield around Earth—as Tony Stark envisioned? Our problem, a problem that confronts many human beings in the MCU, is that life beyond Earth quite possibly will exist in a form we are unfamiliar with, and with unfamiliarity comes fear and anxiety. The extraterrestrial Other is irrevocably monstrous. How is it possible to discern "alien" intentions if we are unable to recognize and understand alternative forms? Often, though not always, when "aliens" arrive on Earth, their intent is either the extermination of humanity or the instrumental and utilitarian use of humans as enslaved workers or breeders. If we wait to see if we are friend, foe, or food, it may be too late.

Unfamiliarity is not always a problem for the characters who inhabit the MCU, as life is remarkably more diverse and more complex, limited only by the imagination of screenwriters and the technical capabilities of make-up artists and digital animators. There, the universe is teeming with myriad life forms and, while most are humanoid in appearance, life in the MCU is not limited to organisms that resemble humans or any other nonhuman Earth-bound organism. And for this reason it is important to take seriously the ethical underpinnings of the superhero genre and how science fictional characters of any and all forms of "life" interact. What if extraterrestrial beings developed their own form of artificial intelligence—a computer microchip, for example—and inserted this into a synthetic body. Would humans recognize this being as a living organism or as an artificial creation? Similarly, would an extraterrestrial being who visits Earth identify computers, robots, and androids as different forms of terrestrial life or as nonliving entities? How we conceive of both the origin and meaning of life therefore mediates our ethical obligations in a posthuman universe—a theme that recurs with remarkable frequency in the Marvel Cinematic Universe, including the primary subject of this chapter, Marvel's *Secret Invasion* (2023).

Although promoted as an espionage thriller, *Secret Invasion* is more properly a science fiction narrative of human-alien encounters and continues the storyline of Captain Marvel. At the conclusion of *Captain Marvel*, both Nick Fury and Carol Danvers (aka Captain Marvel), the Skrull are promised a new home, one where they can live in peace and without fear. In *Secret Invasion*, however, it becomes clear that the Skrull were not provided sanctuary and the narrative arc of the show underscores feelings of betrayal, suspicion, and mistrust between both the Skrull and humans. Despite the show's title, the arrival of the Skrull was neither secret nor invasive. The species came as inter-galactic refugees and their treatment by humans is but one indication of how humanity might respond to a posthuman future populated with

CHAPTER FOUR

extraterrestrials. Before we embark on our exploration of being human in Marvel's *Secret Invasion*, though, it is necessary to provide a more solid foundation of human-alien relationships.

THE HUMAN-ALIEN ENCOUNTER

If we can't protect the Earth, you can be damned well sure we'll avenge it!
—TONY STARK *(AVENGERS* [2012])

In science fiction narratives, including those of the superhero genre, extraterrestrial aliens often perform functions similar to mutants and androids: The figure of the alien is almost always deployed as a metaphor for the Other. As such, the portrayal of human-alien encounters can be either positive or negative, and, in turn, highlight a constellation of emotions among humans, such as fear, mistrust, fascination, suspicion, jealousy, and even love and lust.[10] In most late nineteenth- and early twentieth-century narratives of human-alien encounters, often only a binary solution was possible: eliminate or be eliminated.[11] This trope is readily seen, for example, in H. G. Wells's pioneering story *The War of the Worlds* (1898) and more recent cinematic blockbusters, such as *Independence Day* (1996), *Battle: Los Angeles* (2011), and *Battleship* (2012).[12] These stories, Russell Blackford writes, "are essentially narratives of war and survival, and in most of them humans prevail against the odds." And while these films sometimes engage with questions of morality and ethical choice, Blackford explains, the emphasis is usually on the virtues of courage and innovation. In the superhero genre, including the MCU, alien invaders—notably Thanos—are unquestionably threatening and offer no possibility of compromise. Humans must either violently oppose the threat or succumb to possible extinction. Markedly, though, unlike in films such as *Independence Day* and *Battle: Los Angeles* where ordinary humans are able to directly confront and defeat the alien threat, in the superhero genre humans repeatedly fail—miserably so—to offer any resistance to the extraterrestrial beings and must depend on their more-than-human and other-than-human counterparts for salvation. Indeed, were it not for the heroics of the Avengers in *Avengers: Endgame* (2019), all life on Earth, including that of humanity, would cease to exist.

Rarely do alien invasion narratives provide an opportunity for reflection on the nature of humanity or of humanity's relationship with the cosmic Other. Typically, aliens appear as anonymous threats to humanity. In *Battleground: Los Angeles*, for example, extraterrestrials arrive unexpectedly and begin immediately to wage war against humankind; there is no prospect of dialogue, no opportunity for diplomacy. In the cult-classic *The Blob* (1958), likewise, a jelly-like amorphous species-being lands on Earth and, parasitic-like, starts to consume frightened humans. In this latter film, there is no effort—indeed, there is no opportunity—for humans to communicate with the blob, let alone countenance the being's will-to-live.

That said, many alien invasion stories are haunted by a colonial logics.[13] For example, enslavement, plague, genocide, environmental devastation, and species extinction all figure prominently in science fiction; and yet, John Rieder explains, these are not products of the fevered imaginations of science fiction writers, but instead the bare historical record of what happened to non-European people and lands after being "discovered" by Europeans.[14] The most well-known and influential example of the colonial metaphor and, in many ways, the paragon of the alien invasion trope, is Wells's *The War of the Worlds*.[15] Here, Wells critiques European colonialism through his portrayal of England being invaded by Martians. In this way, Wells uses the invasion trope as a metaphor for Victorian prejudice as well as British imperialism. That said, as Lord cautions, the trope diminishes any possibility for a code of human-nonhuman ethics. She continues that the portrayal of alien invasion "creates a stark moral template of good and evil," and thus obviates any chance of reconfiguring the human-alien encounter along a different ethical axis.[16]

Of course, not all extraterrestrials are portrayed as hostile invaders. So-called 'friendly' aliens are identified primarily through their positive interactions with humans. Unlike the hostile alien who pursues its own interests, often at the expense of their human counterparts, the friendly alien seems to support and to serve human interests.[17] Noteworthy is *E.T. the Extraterrestrial* (1982), a film that centers on a lost and well-meaning alien who is victimized by callous humans.[18] The canonical example, though, remains the 1951 film *The Day the Earth Stood Still*. Set against the backdrop of the Cold War and the nuclear arms race, Klaatu, an extraterrestrial from a technologically superior alien race, arrives on Earth bearing a warning. When he first emerges from his spacecraft, Klaatu is fired upon by US soldiers. In turn, a supersized robot (Gort) emerges from the craft and easily destroys the military's weaponry, thus signaling that humanity is hopelessly outmatched. Klaatu is then taken to a military hospital, where he unsuccessfully tries to explain the purpose of his arrival: that humanity *must* seek peaceful solutions to their on-going conflicts. His words, predictably, fall on deaf ears and Klaatu escapes, to temporarily live among humans, all the while, messiah-like, spreading his warning. Ultimately betrayed by humanity,

Klaatu is again shot, resulting this time in his death. Gort, however, retrieves the corpse and resurrects the other-than-human being. The film ends with a soliloquy delivered by Klaatu. Speaking to all of humanity, Klaatu declares, "The universe grows smaller every day, and the threat of aggression by any group, anywhere, can no longer be tolerated." As such, he counsels, "there must be security for all or no one is secure." In other words, it is necessary that humanity recognize that it is no longer apart from life beyond Earth, and that all species, regardless of origin, are deserving of moral standing and the right to live. The only freedom that will be lost, Klaatu notes, is "the freedom to act irresponsibly." Other inhabitants of the universe have long accepted this principle, Klaatu affirms, explaining further that a "race of robots"—entities such as Gort—was created to police the universe to ensure peace. "In matters of aggression," Klaatu states, "we have given them absolute power. . . . [and] at the first sign of violence, they act automatically against the aggressor." The system is imperfect, Klaatu concedes, but maintains that the result has been universal peace, whereby all species-beings were able to "live in peace, without arms or armies, secure in the knowledge that [all] are free from aggression and war." It was Klaatu's mission—one that was met repeatedly with mortal violence at the hands of humanity—to offer a warning: "It is no concern of ours how you run your own planet, but if you threaten to extend your violence, this Earth of yours will be reduced to a burned-out cinder. Your choice is simple: Join us and live in peace, or pursue your present course and face obliteration. We shall be waiting for your answer. The decision rests with you."[19]

The Day the Earth Stood Still is noteworthy for its attempt to promote a reverence for life. Most films, especially now, sacrifice introspection for violent action. Some films do though confront directly the recognition of the other's will-to-live. In *Arrival* (2016), humans encounter extraterrestrial beings that bear no resemblance to and communicate in a language wholly unfamiliar to humans. A central theme of the film is to understand how to converse with the aliens—a necessary first-step to discern their intentions. As Louise Banks, a linguist and central protagonist of the film explains, "We need to find out, do they make conscious choices or is their motivation so instinctive that they don't understand a 'why' question at all. And biggest of all, we need to have enough vocabulary with them that we understand their answer." Banks, in this scene, evokes something of Schweitzer's reference for life, recognizing that the alien beings have a will-to-live and, thus far, appear to reciprocate. The extraterrestrials demonstrate no overt hostility but instead make gestures suggestive of a common goal: that humans and aliens can learn and communicate with each other without rancor. This required, on

Banks's part, a recognition that her own will-to-live is manifest in every other living species, including those unable to express this will-to-live in ways that we can understand. In other words, we do not need to communicate directly with other species-beings to comprehend *their* will-to-live. It suffices to know that humans are not exceptional in wanting to live a flourishing life.

In modern cinema, the trope of alien (or human) invasions has been supplemented with that of refugee movements.[20] Released in 1988, the film *Alien Nation*, for example, highlights the discrimination and antipathy directed toward the "Newcomers." Approximately 300,000 alien beings crash-land on Earth (near Los Angeles, California), and, in their attempt to integrate into society, face bigotry, hatred, and violence. In *District 9* (2009), likewise, many sick and malnourished extraterrestrial beings also crash on Earth, this time in South Africa, and are confined to a "refugee" camp where they encounter prejudice and animosity from humans. As a metaphor for real world humanitarian crises, Catherine Marshall explains, the film "prompts the viewer to reflect on the ongoing maltreatment not only of the estimated 270,000 registered asylum seekers longing to assimilate into South Africa, but also the 42 million people worldwide who are currently displaced." "If we are incapable of treating our earthly fellows humanely," she wonders, "how can we ever hope to function as morally robust beings in a universe whose boundaries we cannot begin to comprehend?"[21] In other words, if our real world is truly on the verge of becoming posthuman, how will humanity (collectively) respond to more-than-human or other-than-human beings? In science fiction, the alien Other—analogous to mutants and androids—is a metaphorical surrogate for marginalized human beings; when we fail to extend moral standing to the Other, regardless of its species-being, we are primed to commit further atrocities toward alternative life-forms that may simply want to live in peace. To that end, it is noteworthy that films such as *Alien Nation* and *District 9* explicitly frame the human-alien encounter as a chance happening, such as the malfunction of the extraterrestrial spacecraft. In these scenarios, the alien beings do not arrive on Earth by choice and, expressly, harbor no ill-intentions toward humanity. Certainly, if given the chance, the aliens would prefer to return to their own planets. Humans, in turn, can extend moral standing to the interplanetary strangers or—as is usually the case—respond with suspicion and violence.

Above all, fictional human-alien encounters provide much fodder to rethink what it means to be human and our ethical relationship to other beings both on Earth and beyond. Remarkably, though, these questions have long garnered attention in the real world. Indeed, science fiction is often science fact.

CHAPTER FOUR

THE ETHICS OF HUMAN-ALIEN ENCOUNTERS

You've visited other realms, seen different species, aliens. Have you ever
encountered any that were blue?
—PHIL COULSON TO LADY SIF *(AGENTS OF S.H.I.E.L.D.,* "YES MEN,"
SEASON 1, EPISODE 15)

In fictional accounts of human-alien encounters, the meeting often, if not
always, says more about humanity than it does the alien species. When aliens
invade, for example, as they do in *War of the Worlds, Battle: Los Angeles,* and
countless other films, humans appear to set aside (at least, temporarily) their
differences and unite to combat a common enemy; when aliens arrive as
refugees or by accident, as depicted in *Alien Nation* and *District 9* humans are
portrayed (mostly) as xenophobic and without compassion. As such, science
fiction, including the superhero genre, serves as an interesting litmus test to
one of humanity's "great outstanding question of existence," that is, whether
or not we are alone in the universe.[22] Indeed, perhaps since the dawn of
humanity, humans have contemplated the existence of life beyond our planet.
And throughout most of humanity's history, answers were restricted to the
realms of legend, myth, and speculation, informed by religion and philosophy. More recently, however, science has intervened. Beginning in the 1960s,
astronomers using radio telescopes began to search for signals emanating
from space. More recently, the search has expanded, using a variety of techniques and instruments, including the use of unmanned vehicles. In addition, astrobiologists have shifted the search for extraterrestrial life in ways
hardly imagined just decades earlier. In the twentieth century, for example,
scientists on Earth have uncovered living organisms in places long thought
of as lifeless. In fact, as Billings explains, "some forms of life . . . can survive
in virtually all known terrestrial environmental extremes—nuclear radiation, permafrost, temperatures above the boiling point of water, the deep
subsurface Earth, around deep-sea hydrothermal vents, without sunlight,
and so on. Wherever humans or their technological counterparts have gone
on Earth, they have found life."[23] Maybe, some astrobiologists are wondering,
we haven't looked for life in the right places, or in the right form?

The existence of so-called extremophiles on Earth has radically altered our
understanding (and definition) of life itself. As introduced in chapter one, our
scientific knowledge of life has been—and still largely remains—geocentric.
Our search for life beyond Earth until recently was limited to a conception of
life *as we currently know it,* that is, carbon-based cellular life. Armed with the

knowledge that life on Earth appears in places wholly unexpected, it is not far-fetched to speculate that drastically different forms of life might have evolved in ways unique to the decidedly un-Earth-like environments of the trillions of celestial bodies that travel the known universe. It could be, as Louis Irwin and Dirk Schulze-Makuch explain, that the form of life we know is the only one possible; however, there is no logical reason to assume that the laws of physics and chemistry limit the possibilities for life to exist and flourish only in ways with which we are familiar.[24] The challenge, as Christopher Chyba and Kevin Hand note, is that the design of life-detection experiments make explicit or implicit assumptions about what life is, and what measurements will be sufficient to demonstrate its presence or absence.[25] In other words, we might not recognize "life" when we encounter it.

The ethical and moral obligations of humans toward other-than-human life throughout the cosmos is hardly the stuff of science fiction. Since the founding of the United States' National Aeronautics and Space Administration (NASA) in 1958, countless scientists, such as the pioneering luminaries Joshua Lederberg, Melvin Calvin, and J. B. S. Haldane, have wrestled with the consequences of encountering extraterrestrial life.[26] Indeed, in 1959, NASA proposed a project designed to detect evidence of biological activity on Mars, and the following year, Lederberg coined the term *exobiology*, defined as the study of life's origins on Earth and the development of instruments and methods to search for signs of life on other celestial bodies.[27] Admittedly, the early concerns expressed by NASA scientists were largely practical rather than ethical. The potential for scientific research, these individuals worried, would be irrevocably compromised if space exploration was conducted without heed to protect the extraterrestrial environments being explored. Scientists worried, for example, that if terrestrial organisms, such as bacteria, were inadvertently introduced on a distant moon or planet, these could disrupt that celestial body's entire ecology, making subsequent research on pristine extraterrestrial life impossible.[28] That said, scientists were not unaware of the profound ethical and moral consequences involved in space exploration. Subsequently, the field of astrobioethics was founded to consider precisely the extension of moral standing to organisms and/or ecologies beyond the confines of Earth and its environs.[29] In so doing, however, this constellation of scientists confront a deeper question, for to simply ask if there life beyond Earth is to question our most basic understanding of the meaning of life *on* Earth.

A consensus among most astrobiologists is that if extraterrestrial life is found, it is most likely microbial. This begs the question: does extraterrestrial life—even if it appears in microbial form—have moral standing? In other words, do we acknowledge that these beings have their own distinct will-to-live and, consequently, should we affirm a reverence for life? For some scientists, humanity does have an obligation to respect the organisms' will-to-live. "If there is life on Mars," Carl Sagan insisted, "I believe we should do nothing with Mars. Mars then belongs to the Martians, even if the Martians are only microbes."[30] Not everyone in the scientific community is in agreement, however, and this clearly demonstrates that the moral foundations of space colonization continue to exhibit uncomfortable parallels with Earth-bound colonial practices.[31] Dirk Schulze-Makuch and Paul Davies, for example, defend that "colonization from a strictly biological viewpoint is a normal and natural part of the evolutionary process, and as such is ethically neutral." They explain that "organisms on Earth typically explore their environment and colonize new habitats and niches when available, spreading out as must as possible. . . . Any species that doesn't follow this biological imperative is likely to be driven to extinction by the process of natural selection."[32] According to this logic, it is normal and natural for humanity to explore and colonize the cosmos. Such a position, however, is far from being ethically neutral, for it reduces human activity to some preconceived notion of biological instinct. Humans are *impelled* to exploit and occupy interstellar environments out of necessity; through technological advances, we expand the ghastly drama of our mortal existence beyond Earth into the infinite reaches of the universe. Consequently, as Alessandra Marino and Thomas Cheney document, "coloniality remains a fundamental logic of space expansionism and the frontier narrative is a rendition of the deeply engrained fantasies of appropriation of land and resources that push the horizon of possibility always a little bit further."[33]

Despite all efforts to date, however, scientists and astronomers have found nothing but a lifeless universe and an "eerie silence."[34] Humanity's search for, and inability to find, fellow travelers in the universe, however, is not without meaning. As Billings writes, the search for life elsewhere has affected the way we think about our home planet and the life on it.[35] Consequently, even the absence of life beyond Earth holds tremendous meaning for many people, for our lonely existence seems to confirm humanity's exceptionalism. Remarkably, the seemingly lack of life throughout the universe—as we saw earlier—has compelled some billionaires, such as Elon Musk and Jeff Bezos—to redouble their efforts to colonize space, if only to promote and protect the "gift" of humanity beyond Earth.

In the fictional world the universe is teeming with alternative forms of life and the Marvel Cinematic Universe is no exception. The Kree, for example, appear frequently in the MCU and figure prominently in the *Guardians of the Galaxy* trilogy, *Captain Marvel, The Marvels,* and the television series *Agents*

In *Agents of S.H.I.E.L.D.* (2018), Kasius (Dominic Rains) and Sinara (Florence Faivre) are Kree, a humanoid species from the planet Hala.

of S.H.I.E.L.D. The Kree are human in appearance, although some individuals have pointed ears. The most distinguishing trait, however, that visually marks the Kree as other-than-human is their (usually) blue skin. Notably, though, several Kree (known as 'pink' Kree) have Caucasian skin-tones and are thus able to easily pass as humans. The Skrull, on the other hand, are humanoid in appearance, that is, they have one head, two arms, and two legs. However, in their natural state, they are reptile-like, with green, scaly skin, furrowed chins, and pointed ears. This is significant insofar as "malevolent aliens are often shown as physically monstrous, typically with reptilian or insectoid characteristics."[36] Consequently, the physical appearance of the Skrull immediately situates them as the monstrous Other. In addition, the Skrull are shapeshifters, meaning that they can take on the appearance of other species-beings, including humans. This ability contributes to the suspicion harbored by many humans that the "alien" is among us, passing unseen and infiltrating society for nefarious purposes.

From humanoid species, like the Skrull and Kree, to large, tentacled beings such as the abilisk, humans interact with a diversity of life far beyond any expectations real-world scientists could ever imagine. As such, the MCU operates in a way similar to our myths and legends of the past: as an opportunity to ponder humanity's meaning in a more-than-human universe. And if the first thirty-plus films and over twenty television shows are any indication, we have cause for concern. Davies is correct to note that humanity's uniqueness in the universe is an enduring question. More troubling, I think, is the possible response of humans on learning that we're not alone. If we discover that life exists beyond Earth, how will humanity respond? The answer to this question is most pressing when understood in the context of space settlement. When European settlers, for example, colonized the New World, their hubris denied the humanity of Indigenous populations and violently took possession of lands deemed *terra nullius*. Does a similar fate await other-than-human beings that currently inhabit the celestial realm beyond Earth? If we truly become a multiplanetary species, we will manifestly become an extraterrestrial invasive species.

SECRET AND NOT-SO-SECRET INVASIONS

> I invite you to join me in the extinction of the human race.
> —GRAVIK (*SECRET INVASION*, SEASON 1, EPISODE 3, "BETRAYED")

> What I want is for you to stand down and
> stop murdering innocent humans.
> —TALOS (*SECRET INVASION*, "BETRAYED")

The series *Secret Invasion* (2023) has not received much critical commentary, perhaps in part because the show made little imprint on Marvel's fanbase. Although widely anticipated—the series would be the first to focus near-exclusively on fan-favorite Nick Fury—*Secret Invasion* underwhelmed audiences and critics alike. Matt Purslow, for example, describes the series as "an underwhelming and often dull espionage show" that "fails to capture the excitement and threat that its *Invasion of the Body Snatcher's*-like plot chases."[37] Despite tepid reviews, however, *Secret Invasion* is an important part of the MCU, insofar as it confronts head-on humanity's reception of other-than-human beings that seek nothing more than a place of refuge. The plight of the Skrulls—the extraterrestrial species at the heart of the show—is made all the more poignant when set against the reception of another species of intergalactic refugees, the Asgardians.

To fully appreciate the plot of *Secret Invasion* it is necessary to review another controversial film, *Captain Marvel* (2019). Notably, this earlier film marks the first appearance of Carol Danvers (Captain Marvel), and the film is the first Marvel film to cast a female superhero as its lead. For this alone, the film has garnered considerable attention and has generated much scholarly critique.[38] Much of this scholarship has rightly focused on the portrayal (and negative fan reception) of Captain Marvel from a feminist lens. Less focus, however, has addressed the film within the context of human-alien encounters. This is a crucial lacunae, in that *Captain Marvel* is also the first film within the MCU that openly confronts humanity's response to extraterrestrial beings. Certainly, otherworldly beings appear in earlier films: Thor, who makes his appearance in *Thor* (2011) is plainly extraterrestrial; and the Avengers (with Thor) battle the Chitauri, a hostile alien species in *The Avengers* (2012). In both of these films, however, the human-alien encounter is muted. In the former film, Thor is treated as anything but an extraterrestrial and, in the latter film, the invading aliens appear as nothing more than cardboard characters, devoid of agency or presence. *Captain Marvel* is different in that the film explicitly centers on the arrival of, and humanity's response to, extraterrestrials on Earth. Indeed, a number of aliens, notably Talos, a Skrull, and Yon-Rogg and Mar-Vell, both Kree, figure prominently in the film's narrative.

Captain Marvel opens with Vers, a member of an elite Kree military taskforce known as Starforce. Vers, however, suffers from amnesia and is plagued by recurring nightmares that include a mysterious woman. Vers is exceptionally skilled and demonstrates enhanced powers which she cannot fully command. Vers's superior, Yon-Rogg, mentors the young fighter in her training and counsels her to control her emotions and dismiss the nightmares. At one point, Yon-Rogg exclaims, "I'm so proud of you. You've come a long way since that day I found you by the lake. But can you keep your emotions in check long enough to take me on? Or will they get the better of you as always?"[39]

During a mission to rescue Soh-Larr, an undercover operative on Torfa, a neighboring planet, Vers is captured by the Skrull. While in captivity, a Skrull leader named Talos—for unknown reasons—attempts to tap into Vers's mind to recover some undisclosed information. Vers escapes, however, and crash lands on Earth, pursued by Talos and several other Skrull. Vers *looks* human, with flowing blonde hair and brown eyes. Her clothes and technology, though, are decidedly futuristic and her arrival captures the attention of Nick Fury and S.H.I.E.L.D. Initially, Fury suspects that Vers is delusional but, after witnessing firsthand the shape-shifting capabilities of the Skrull—and seeing them in their normal reptilian form—Fury determines

In *Captain Marvel* (2019), Carol Danvers (Brie Larson), Maria Rambeau (Lashana Lynch), Nick Fury (Samuel L. Jackson) learn from Talos (Ben Mendelsohn) that the Kree are committing genocide against the Skrull.

that Earth is under threat from an alien invasion. With the help of Fury, who has access to secret records, Vers learns that she is not Kree but human. Adding to the mystery is that Vers supposedly died on Earth several years earlier when her plane, copiloted by the mysterious woman in Vers's nightmares, crashed near a lake. Armed with this information, Vers and Fury track down Maria Rambeau, a former US Air Force pilot. Rambeau explains that Vers's real name is Carol Danvers and that she was a highly trained pilot in the US Air Force. At this point, Talos arrives, having tracked Fury and Danvers to Rambeau's home. Together, Danvers, Rambeau, and Talos are able to reconstruct the past. The mysterious woman was Danvers's former mentor and Commander, Wendy Lawson. However, Lawson was working undercover in the US Air Force. Her true name and identity was Mar-Vell, a Kree scientist who was developing a light-speed engine.

Six years earlier, Danvers and Lawson were test-piloting the prototype engine when their plane was shot down by an alien craft—piloted by Yon-Rogg. Gravely injured, Lawson explains to Danvers that the Skrull refused to submit to the Kree and become a colonized people. In retaliation, the Kree waged a genocidal campaign against the Skrull. Mar-Vell empathized with the Skrull and saw in them nothing more than a will-to-live, bemoaning that she "spent half [her] life fighting a shameful war." In the film's most prominent plot twist, Danvers (and Fury) realize that the reptilian Skrull, and not the human-like Kree, are the "friendly" aliens. The Skrull in fact are inter-galactic refugees, fleeing persecution from another alien species, the Kree. Here, Mar-Vell seemingly captures the ethos of Albert Schweitzer, whereby "true philosophy must start from the most immediate and comprehensive fact of consciousness, which says:

Tyler Hayward (Josh Stamberg) is willing to protect humanity at all costs in *WandaVision* (2021).

'I am life which wills to live, in the midst of life which wills to live.'[40] Mar-Vell sees in the Skrull not an enemy species, but instead a species that wants nothing more than to live and to flourish.

Determined to make amends, Mar-Vell hopes that the light-speed engine will help the surviving Skrull reunite and escape to a distant planet far from the murderous reach of the Kree. It was during a test run of the engine that Yon-Rogg shot down the plane. Hoping to acquire the technology for the Kree, Yon-Rogg confronts the two and kills Mar-Vell. Danvers then attempts to destroy the engine by shooting it with a Kree pistol given to her by Mar-Vell. In the subsequent explosion, however, Danvers is exposed to the energy-core of the engine, thereby acquiring her enhanced powers. Rendered unconscious, Danvers is taken by Yon-Rogg to his home planet of Hala where she is later weaponized in the war against the Skrull. Having amnesia, Danvers is easily manipulated into believing that she is a Kree fighter.

Following these events, now with the presence of life beyond Earth established, the US government in 1995 redirects its space program with the creation of the Sentient Weapon Observation and Response Division, known simply as SWORD. Initially led by Rambeau, SWORD's purpose was primarily intelligence gathering. Following a series of events, including the Chitauri invasion of the Battle of New York and the fight against Thanos in the Battle of Earth, the function of SWORD changed dramatically. This was brought to bear under the directorship of Tyler Hayward, a heartless and driven man who displays a strong animosity toward more-than-human and other-than-human beings.

After the death of Rambeau, who succumbed to cancer, Hayward was appointed as acting director of SWORD. As part of Project Cataract, Hayward

Nick Fury (Samuel L. Jackson) enlists Gravik, a young Skrull refugee, to conduct covert operations in *Secret Invasion* (2023).

has the body of Vision (who was killed by Thanos in *Avengers: Infinity War*) taken to a secret facility. Developed in secret, the purpose of Cataract was to disassemble and reassemble the synthetic being to create a sentient weapon or, in the words of FBI Agent Jimmy Woo, "to bring him back online."[41] When confronted by Wanda Maximoff (Scarlet Witch), Vision's long-time paramour, Hayward concedes, "We're dismantling the most sophisticated, sentient weapon ever made." Subsequently, when Maximoff decries "But Vision's not a weapon. You can't do this," Hayward counters, "In fact, it is our legal and ethical obligation."[42] Likewise, when Monica Rambeau, the daughter of Maria and now SWORD agent questions Hayward on the ethics of weaponizing Vision's body, the acting director is equally dismissive. SWORD, Hayward explains, "shifted away from manned missions and refocused on robotics, nanotech, A.I. sentient weapons, like it says on the door." "It also says," Monica counters, "'observation and response' on that door, not 'creation.'" Hayward replies, simply, "World's not the same as you left it. Space is now full of unexpected threats."[43] Markedly, many of those unexpected threats come about because of the hubris of humans, including the one-time director of S.H.I.E.L.D., Nick Fury. It should not go unmentioned, also, that Hayward's oversight and justification of the Cataract project mirrors that of the former NAZI and HYDRA leader Werner Reinhardt/Daniel Whitehall and his medical experiments on Inhumans.

In 1997, Fury meets with several high-ranking members of the Skrull, including the more senior Talos, Talos's daughter, G'iah, and a younger, headstrong Skrull named Gravik. Orphaned when his parents were killed by the Kree, Gravik is intelligent, resourceful, and resilient.[44] Pointedly, the Skrull simply want to establish a place of refuge, safe from the Kree. However, Fury believes—not without some justification—that the Skrull would never be able to fully integrate into human society. Reptilian in their natural appearance, they are only able to appear in public by shapeshifting into human form. This negation of their self-identity is deeply humiliating for the Skrull, who want nothing less than to "live in their own skins."[45] As Fury concludes, "There is not enough room or tolerance on this planet for another species!"[46]

Conceding humanity's inability to demonstrate any reverence for life beyond their own species, Fury promises to find the Skrull refuge beyond the confines of Earth. However, his altruism is tempered by a more pragmatic attitude, namely to weaponize the Skrull in Fury's self-appointed mission to defend Earth from other nonhuman beings. For Fury, the relationship is one of quid pro quo. As he promises the Skrull, "You keep your word and I'll keep mine."[47]

Although the Skrull use their shape-shifting abilities solely for defensive purposes, Fury, always the operative, sees an opportunity to cultivate the persecuted refugees as super spies. Unlike Inhumans, however, the Skrull were never meant to be actual members of S.H.I.E.L.D., operating more as Fury's own private spy network.[48] Indeed, Fury enlisted the Skrull to engage in a number of covert operations, including targeted assassinations. This is revealed years later when Gravik reminds Fury of his first mission. "You know," Gravik accuses Fury, "I was young. I tried my best." He continues, "I wanted to give everything to impress my hero. You know, the one who promised us a home. But this man. . . . This man. Fury, he had a wife. And he had children. Yeah, all right, maybe he was a bit misguided. But I killed him because of you. I killed him and I killed so many more. And everyone I killed took a little piece out of my heart."[49] Here, Gravik's accusation is also an admission that he was psychologically scarred and struggled with the guilt by his participation in Fury's ghastly drama of extrajudicial killings. More pointedly, Gravik resents Fury for his apparent betrayal in finding the Skrull a home. It is this feeling of duplicity that compels Gravik to disassociate from the seemingly more biddable Skrull who continue to place their faith in humanity. As Maria Hill, who is later killed by Gravik, describes in her assessment of Gravik, he "preys on the collective rage of young, displaced Skrulls."[50]

A natural-born leader, Gravik assembles a renegade faction of refugees to form the Skrull Resistance. For Gravik, humanity—personified by Fury—betrayed the trust of his species and, while failing to assist the Skrulls in finding a place of refuge, actually placed them in greater danger as secret agents. No longer consoled by Fury's platitudes, Gravik devises an elaborate plan to wrest control of Earth from humanity. The first step is to conduct a number of false flag operations—similar to the tactics used by

Maria Hill (Cobie Smulders) is distrustful of Gravik in *Secret Invasion* (2023).

the Watchdogs—to create havoc and fear. However, Gravik hopes to capitalize on the nationalist prejudices humans harbor toward other humans, notably the geopolitical tensions between the United States and Russia. Operating behind the fictitious "Americans Against Russia" movement, the Skrull intend to foment a series of terrorist attacks whereby the Americans will blame Russia and the Russians will blame the Americans. As Gravik explains, "Humans will be at all-out war with each other within the week. And while they're at each other's throats, we're gonna break their backs."[51] Ultimately, if all goes according to plan, the conflict would escalate to a nuclear conflagration between the two superpowers—a derivative plot device that appears in several novels and films, most notably the Terminator franchise. Crucially, Skrull are immune to the harmful effects of radiation. As such, after the last bombs are detonated and all of humanity is exterminated from nuclear fallout, the Skrull will remain as the most powerful, intelligent species on Earth. And should the Avengers attempt to stop the Skrull, Gravik engineers a super soldier program using DNA of superheroes, including Carol Danvers. As Gravik explains, "The only way we can . . . claim this planet as our home is to become super ourselves. Now, we no longer just change faces. We change powers. We're gonna be uniquely programmed weapons of mass destruction. All of us. Super Skrulls."[52]

Fury only belatedly learns of the "secret invasion" planned by Gravik and the renegade Skrull. For years, in fact, Fury was off-Earth, developing a complex aerospace defense system known as the Strategic Aerospace Biophysics and Exolinguistic Response, or SABER. Once aware of the plot, however, Fury's efforts to stop Gravik are hindered because many government officials

Gravik (Kingsley Ben-Adir), in human form, organizes a Skrull Resistance to wrest control of Earth from humans in *Secret Invasion* (2023).

are in fact rebel Skrulls posing as humans. In an ironic twist, Fury's previous betrayals of the extraterrestrials has come back to haunt him.

Following the false-flag operation that resulted in significant loss of (human) life, several Skrull condemn Gravik for his actions. "You murdered over two thousand innocent humans. Children," a senior Skrull accuses Gravik, "and yet, you don't seem remotely remorseful."[53] And Gravik, similar to Ultron's take on humans, is in fact dismissive of the carnage he has wrought and harbors no regrets. "Long before we arrived on this planet," Gravik retorts, "[humans] were destined to consume themselves. So, for anyone flinching at the thought of innocent deaths, let me assure you, we're only hastening the inevitable." This, also, is a recurrent theme of the MCU. It is not simply that humans commit violence against mutants, androids, and aliens; humans regularly commit violence against their own species and show little reverence for life of any kind. Later, when Gravik is compared to a dog, the rebel leader concedes that he much prefers being in the company of dogs instead of humans: "I quite like dogs. In fact, I prefer them. Dogs aren't hypocrites. And they don't lie. They don't lock each other up in cages. They don't pimp, poison, and they don't go out of their way to degrade and destroy their own habitat."[54] It goes without saying that Gravik's charges *are* bolstered by the Skrull's uneasy history with humans. As Erik Amaya writes, "It is easy to sympathize with the Skrull insurgency when you consider they waited thirty-plus years for Fury and Carol Danvers to find them a new home world, and Fury used their abilities to assemble an undetectable spy network/execution squad in the interim."[55]

In a revealing scene, Gravik asks Talos if he knows "the difference between statesmen and soldiers." Gravik answers his own rhetorical question, explaining that "one lot spends the war posing for pictures, while the other does all the killing and the dying."[56] In this instance, however, it is not simply a dichotomy between soldiers and statesmen, but instead between humans and other-than-human beings. Fury's manipulation of the Skrull is toward the supposed betterment of humanity—or at least toward those sections of humanity Fury deems proper. The Skrulls are rendered little more than useful pawns, sacrificed when necessary. If Gravik is a ruthless, compassionless killer, it is because of the trauma induced by becoming weaponized in the service of humanity. Still, Talos attempts to reason with Gravik. "You're gonna take our people to the edge of extinction with the war with the humans," Talos warns. Undeterred, Gravik responds simply, "All these miscreants know is murder. Look how they treat each other." Notably, Talos doesn't disagree and, in fact, seemingly concedes the violent tendencies of humanity.[57]

In the series finale, Gravik apparently confronts Fury: "You pimped us, Fury. You put us out to work for you and when you were done with us, you threw us away. So, first, I'm gonna kill you. And then I'm gonna take a flamethrower to humanity."[58] However, the man standing opposite Gravik is not Fury, but instead G'iah, a Skrull who shapeshifted into Fury's form. This scenario allows the now-enhanced G'iah to engage in mortal combat with Gravik, also superpowered, but the climactic battle also deflects and detracts from Fury's own reckoning. Humanity—embodied by Fury—is once again absolved of its violent subterfuge and two members of the same species battle to the death to determine the fate of Earth. Ironically, the series ends not with a conflagration between two geopolitical superpowers (US and Russia) but instead between two superpowered aliens. Whereas Gravik hoped to turn humans against each other in a final showdown, the series pinnacles with Skrull fighting Skrull. The fate of humanity is secured, but at what cost? The Skrull, fleeing genocidal persecution as intergalactic refugees, endure prejudice and discrimination on Earth, and then face betrayal and manipulation, and ultimately are a fractured species, divided amongst themselves. Moreover, when the scale and scope of the Skrull Rebellion is finally revealed, US president Ritson calls on Congress to pass an emergency authorization bill that would designate all off-world foreign species as "enemy combatants." He affirms: "We know how to find you and we will kill every last one of you." Despite being saved by a mortally wounded Skrull just days earlier, Ritson's extraterrestrial xenophobia shows no bounds as he sanctions the extrajudicial execution of Skrulls. Not surprisingly, the biopolitical rhetoric of Ritson fuels public fear and sparks the formation of numerous vigilante groups intent on exterminating other-than-human beings. This violence, similar to that meted toward Inhumans in *Agents of S.H.I.E.L.D.*, quickly devolves into mass hysteria, as humans indiscriminately harm and kill *anyone* perceived as different. As Fury laments, "Now you have dumb ass vigilantes killing humans too."[59]

Upon their arrival, the Skrull harbored no plans to conquer Earth but instead came in peace, hoping only to fit into human society until at some point they were able to find another planet suitable for permanent settlement. As intergalactic refugees fleeing the genocidal Kree, the Skrull, not unlike the Inhumans at Afterlife, hoped for nothing more than a sanctuary, a place to be free from persecution. In *Captain Marvel*, Talos makes a plea to humanity, personified by Fury and Danvers: "My people lived as refugees on Torfa. Homeless, ever since we resisted Kree rule and they destroyed our planet. Now the handful of us that are left will be slaughtered next. Unless you will help me finish what Mar-Vell started. . . . We just want a home."[60] As the monstrous Other, however, the Skrull are subject to prejudice and discrimination. Fury tries to defend his inability (or unwillingness) to provide refuge for the Skrull. "It's easier to save the lives of eight billion people," Fury concedes, "than it is to change their hearts and minds."[61] Here, though, Fury is disingenuousness, for humanity has welcomed with open arms other extraterrestrial aliens. As such, the treatment of the Skrull refugees stands in marked contrast with that of Thor and the Asgardians.

In *Thor* (2011), the titular character arrives on Earth, some twenty years after the appearance of the Skrull and the Kree, as depicted in *Captain Marvel*.[62] Thor is extraterrestrial, and yet Thor is also the personification of white hypermasculinity, hence his presence undermines conventional readings of human-alien encounters. It is curious, therefore, that many commentators focus on the question of Thor's god-like status and sidestep altogether the obvious fact that Thor is undeniably not of Earth. In his physical appearance, there is nothing "alien" about Thor nor, for that matter, is there anything remotely alien about most Asgardians (although some appear to have pointed ears). Indeed, the human characters who initially meet Thor—notably Jane Foster and Erik Selvig—are particularly nonchalant in their attitudes toward the extraterrestrial. Perhaps it is not surprising, therefore, that Thor—the physical embodiment of the white male hero—is soon accepted into the ranks of humanity. Selvig takes Thor drinking and Foster become romantically involved with the extraterrestrial. Likewise, a common purpose is acknowledged that unites Thor

Despite being an extraterrestrial, Thor (Chris Hemsworth) is welcomed by Darcy Lewis (Kat Dennings), Erik Selvig (Stellan Skarsgård), and Jane Foster (Natalie Portman) in *Thor* (2011).

and S.H.I.E.L.D., personified by Phil Coulson. "You and I," Thor affirms, "we fight for the same cause: the protection of this world. From this day forward, you can count me in as your ally." This declarative statement marks the beginning of Thor's acceptance, and that of the Asgardians, into humanity's moral community. S.H.I.E.L.D.'s embrace of the Asgardians highlights even more Fury's betrayal of the Skrull, who for years engaged in covert operations in defense of humanity.

When Asgard is destroyed by Thor's sister, Hela, in *Thor: Ragnarok*, the Asgardians are forced to seek refuge elsewhere in the universe. Later, in *Infinity War*, their transport vessels are attacked by Thanos, who slaughters many Asgardians.[63] With few options, the surviving Asgardians arrive on Earth as a people without a home. Unlike the Skrull, though, the Asgardians are encouraged to integrate into human society and establish a settlement in Norway known as New Asgard.[64] Remarkably, New Asgard even becomes something of a tourist destination, complete with a family-oriented theme park where human children can frolic and play among the extraterrestrials. The reception of the Asgardians reveals in full the hypocrisy of Fury and those who sided with him. When the Skrull asked for help, Fury condemned humanity, affirming that humans would never accept or even tolerate the presence of extraterrestrial beings on Earth. And yet, when the white-coded Asgardians sought refuge, they are rendered political innocents and not subjected to the same treatment as the Skrull.[65] It seems that humanity's longstanding racisms and bigotry can be universally applied.[66]

CONCLUSIONS

In the realm of science fiction, dreams of traveling to distant planets and even of colonizing space are plentiful.[67] From the mid-twentieth century onward, the dream of human beings traveling beyond the confines of Earth has become a reality, and the fantasy of humans settling distant cosmic bodies has entered the realm of possibility. However, the hope to colonize outer space is not shared by everyone. Indeed, for every argument in favor of space colonization there is an argument against such an undertaking.[68] Before we address these debates, however, it is necessary to provide a basic foundation, namely the purported rationale for dispersing humanity throughout the universe.

Broadly, there are five basic justifications for the colonization of space: human survival; resource base expansion; scientific investigation; adventure; and spiritual growth.[69] Of these reasons, I am concerned most with those centering on human survival. Indeed, perhaps the most enduring justification for space settlement is that it is a necessary means of securing the long-term survival of the human species in the face of possible global catastrophes—a justification, we should note, forwarded by the fictional Skrull.[70] This is certainly the position of people like Martin Rees, who, writing in 2003, states that "the odds are no better than fifty-fifty that our present civilization on Earth will survive to the end of the present century." He explains: "Our choices and actions could ensure the perpetual future of life (not just on Earth, but perhaps far beyond it, too). Or in contrast, through malign intent, or through misadventure, twenty-first-century technology could jeopardize life's potential, foreclosing its human and posthuman future. What happens here on Earth, in this century, could conceivably make the difference between a near eternity filled with ever more complex and subtle forms of life and one filled with nothing but base matter."[71] The more pressing danger, though, for Rees and others, is that of complacency. The transhumanist Nick Bostrom, for example, counsels that while humanity has survived "natural" existential risks for hundreds of thousands of years—floods, earthquakes, and the like—humanity "is introducing entirely new kinds of existential risk—threats we have no track record of surviving." As such, Bostrom concludes, "Our longevity as a species. . . . offers no strong prior grounds for confident optimism."[72]

It is important to understand why, well into the twenty-first century, the sentiments expressed by Rees and Bostrom, among others, have found widespread acceptance within the scientific community and elsewhere.[73] In part, the answer is straightforward, as Matt Boyd and Nick Wilson write:

"A number of existential threats to the survival of humanity or human civilization exist."[74] Some are "natural" and include super-volcanic eruptions, world-shattering earthquakes, asteroid impacts, and pandemics; the more daunting, however, are of recent vintage, born of human technology: nuclear Armageddon, artificial intelligence, nanotechnology, and geoengineering and/or bioengineering failures. Faced with such a range of threats, it is hardly any wonder why a chorus of people are warning that humanity is at serious risk of extinction. And on that note, it is no coincidence that in the fictional realm of the MCU, superheroes repeatedly confront extinction-level events, from Ultron's robot revolution to Dormammu's attempt to destroy Earth to Thanos's effort to exterminate half of all living things throughout the universe.

Contemporary fears of humanity's extinction are underscored by humanity's hubris, an arrogance of anthropocentric thinking that attempts to *preserve* both *Homo sapiens* as a species and also the supposed values that separate humanity from all other beings. Accordingly, in recent years, research centers committed to the study of existential risks to humanity have proliferated, including the Future for Humanity Institute at the University of Oxford, the Center for the Study of Existential Risk at the University of Cambridge, and the Center for Security and Emerging Technology at Georgetown University.[75]

In *The Avengers: Age of Ultron* (2015), billionaire Tony Stark hopes to design a protective shield—a suit of armor—around the Earth to defend (primarily) humanity against future alien invasions. His fears are not unwarranted; three years prior, Stark and the other Avengers battled a species of extraterrestrial beings known as the Chitauri, marking the first large-scale violent confrontation between humanity and aliens. As Stark recounts, "that's the endgame."[76] In the real world, billionaires such as Musk, Bezos, and Richard Branson—and their legions of scientists, engineers, and pundits—are interested less in defending Earth than they are in colonizing other planets.[77] Musk, for example, predicts that "history is going to bifurcate along two directions." One path, he argues, is to stay on Earth forever and confront "some eventual extinction event," while the other "is to become a space-bearing civilization and a multi-planetary species."[78] In fact, Musk has repeatedly declared that the long-term survival of humanity can only be guaranteed by spreading out into the cosmos.[79]

The idea that space exploration and colonization of other planets and moons is necessary to preserve the human species is a common trope among many futurists and transhumanists.[80] If humanity disappears, the argument follows, so too does morality itself. Hans Jonas, for example, warned in the 1980s that humans alone are responsible for their continued survival;

indeed, the preservation of humanity and of the entirety of a moral universe is our "cosmic responsibility."[81] In other words, Jonas writes, "There is an unconditional duty for mankind to exist."[82] The human exceptionalism of space colonization is plain to see: Humans are unique among all species as moral beings. It is imperative, therefore, that humans become guardians of the galaxy to ensure that the cosmos is forever guided by our beneficence.

To best fulfill our cosmic responsibility, both futurists and transhumanists promote the notion that humanity requires a "backup planet." The logic is straightforward. "The idea," Ryan Gunderson and his coauthors explain, "is that humans must find ways to live off-Earth, as our planet is becoming increasingly difficult to live on and may become impossible to live on as we face escalating existential threats."[83] Accordingly, Musk supports the colonization of celestial bodies, notably Mars, while Bezos and Branson promise a future world of floating space cities.[84] Not surprisingly, these billionaires and many others, including Bill Gates and Jaan Tallinn, have invested significant resources into countless think-tanks and institutes devoted to existential risks.[85]

Given the specter of extinction-level events foreshadowed by futurists and transhumanists, space colonization becomes an *anticipatory action*, that is, "A seemingly paradoxical process whereby a future becomes cause and justification for some form of action in the here and now."[86] In this instance, an uncertain and indeterminate future provides the grounding for the possible conquest of celestial bodies in our solar system and beyond. That said, the argument that humans need to colonize other planets and moons because Earth may not be habitable for us at some point avoids consideration of the need to keep our planet habitable for the billions of people who will not have the opportunity to leave and thus survive.[87] Indeed, it is not clear *who* will be able to participate in any possible colonizing endeavor. As Schwartz and others caution, "If human survival is among our reasons for pursuing space settlement, then we must think carefully about what we mean by 'preserving humanity.'"[88] This is a pressing question, more so when we expand the question to account for a posthuman future. This theme, incidentally, is vividly portrayed in the satirical science fiction film *Don't Look Up* (2021). The harsh reality is that any settlement scheme will be small-scale; the vast majority of humans—and most nonhuman species—will remain fixed on Earth.[89] Moreover, as Lynda Williams cautions, "Seeking space-based solutions to our earthly problems may actually exacerbate the planetary threats we face."[90]

Still, most proponents and supporters of space colonization remain nonplussed. Milan Ćirković, for example, downplays arguments centered on social justice, claiming that such concerns are "hardly applicable to the specific issue of human space settlement."[91] Indeed, Ćirković asserts, somewhat incredulously, "It is exactly the poor and the tired who are likely to benefit the most from space settlement and the immense prosperity stemming from the expansion of human creativity and material resource base, *as it was the case throughout human history.*"[92] One wonders which world Ćirković has in mind in making this assertion.

Ultimately, as Szocik explains, "At the heart of the arguments skeptical of space settlement. . . . is the fear that the project will exclude already marginalized and excluded groups of people." In fact, Szocik underscores that "this concern is similar to that of the fight against climate change" in that "adaptations to climate change are more easily implemented in rich countries and for wealthy citizens. . . . while poor people in the Global South in particular, but also traditionally marginalized groups in rich countries experience the negative effects of climate change to a greater extent."[93] Echoing these sentiments, John Traphagan explains, "One need not be a rocket scientist to see that for a very large proportion of the world's population, space exploration is likely to represent an experientially remote or even meaningless activity pursued by people who live in a world quite distant, even alien, from their own— where most activities are focused on figuring out how to survive on a daily basis. Much of the world's public are concerned with finding their next meal or securing shelter that is at least marginally livable."[94] On this point, Linda Billings rightly asks, "How many poverty-stricken Bangladeshis, how many sub-Saharan Africans, how many permanently displaced Syrian refugees, how many disabled and unemployable workers could come up with $200,000—or $2,000, for that matter—to move to another planet and start a new life? What are the ethics of giving the rich yet another advantage over the poor?"

Secret Invasion inverts the narrative of space colonization. The Skrull—as a species-being with a will-to-live—confront a genocidal threat and their possible extinction. Their home planet has been rendered uninhabitable and they are forced to seek refuge elsewhere in the universe. And similar to another displaced species-being—the Asgardians—the Skrull do not arrive on Earth as colonizers, but instead as an oppressed people in need. However, whereas the very human-like Asgardians are welcomed and soon integrated into humanity, the seemingly monstrous Skrull, seeking nothing more than asylum, are shunned and ultimately betrayed. Far from a conventional alien invasion story, *Secret Invasion* is an indictment of humanity and the proclivity to turn our heads away from people deemed less-than-human, people who want nothing more than to live a meaningful life free from persecution.

Chapter Five

ENDGAMES

Humans only live to fifty? What's the point of even being born?
—MANTIS *(GUARDIANS OF THE GALAXY, VOL. 3* [2023])

Life isn't meant to be lived alone.
—MELINDA MAY *(AGENTS OF S.H.I.E.L.D.,* "MISSING PIECES," SEASON 6, EPISODE 1)

It is not uncommon for people to complain that their lives lack meaning; they yearn for *something* to give them purpose. In *Agents of S.H.I.E.L.D.*, Alphonso "Mack" Mackenzie finds meaning in living a simple life. "Simple," Mackenzie elaborates, "is how you live a good life. Not with your theories or prophecies. It's following the good word and doing the right thing every time. It's as simple as that."[1] For most of us, finding meaning in life is not so straightforward. "When people ask about the meaning of life," John Fischer explains, "they are not typically wondering about the purpose of human life in general. Rather, they are asking about finding (and experiencing) meaning in being human."[2] To this end, questions about *meaning in life* direct attention toward attendant concepts such as nihilism, suffering, misery, morality, and—markedly—death itself.[3] Indeed, our ability to find meaning in life is necessarily conditioned by our awareness of death, that is, we are aware, either explicitly or implicitly, of life's finitude, our own "endgame."[4]

For many people, death is "bad" in part because it *deprives* us of the ability to pursue projects that give our lives meaning. On this point, as John Fischer elaborates, the "badness" of death is intimately connected to meaningfulness in life because it ends those pursuits that help give meaning to our lives.[5] Martha Nussbaum states, "The intensity and dedication with which very many human activities are pursued cannot be explained without reference

In *Agents of S.H.I.E.L.D.* (2013–2020), Alphonso Mackenzie (Henry Simmons) stands in as humanity's moral compass.

to the awareness that our opportunities are finite."[6] This sentiment is tenderly conveyed when Phil Coulson speaks of his pending death. Faced with a terminal illness, Coulson is asked by Mack about the hardest part of dying. Without hesitation, Coulson responds: "The life I have yet to live."[7]

The human species-beings' *knowledge of the certainty of death* and its necessity to live with the constant awareness of that fact conditions the social and cultural organization of all known human societies.[8] Truly, as Olivia Stevenson and colleagues observe, "Death and bereavement are inevitably central to everyday life across culture, although associated social and cultural practices vary by time and place, ranging from sequestered social taboo to an integral part of ongoing social life."[9] For Zygmunt Bauman, "Once learned, knowledge that death may not be escaped cannot be forgotten—it can only not be thought about for a while, with attention shifting to other concerns."[10] In other words, human life is an ongoing process of learning to live with death.[11] This is seen in our practices as they relate to the dying and the dead, for example, in the ways we grieve, mourn, and remember the dead.[12]

An awareness of death leads naturally to a consideration of other, more metaphysical questions, such as the existence of God, spirituality, and the afterlife. Indeed, as human beings, many of our beliefs about living meaningful lives are conditioned by attendant beliefs about what happens after death occurs.[13] In various Christian accounts, for example, one's immortal

Jemma Simmons (Elizabeth Henstridge) and Leo Fitz (Iain De Caestecker) often question the meaning of life and death in *Agents of S.H.I.E.L.D.* (2013–2020).

soul ascends to heaven and thereupon dwells in eternal peace.[14] These interpretations, among others, speak to the deep human desire felt by many that our worldly existence should have some ultimate meaning or purpose or significance.[15] By way of contrast, existentialist philosophers such as Søren Kierkegaard, Jean-Paul Sartre, and Albert Camus denied the existence of an afterlife. Consequently, death itself lends no meaning to life.[16] More recently, scholars have considered how technological advances, especially in the fields of medicine and healthcare, have altered conceptions of death, to the extent that death is not only something undesirable, but something to be mastered and overcome.[17] Regardless of one's attitude toward death, be it fear or ambivalence, something to be welcomed or avoided, or of the existence of an afterlife in whatever form, the observation holds that humans are *aware* of death and the implications that such finitude holds for living after the death of another. As Jemma Simmons explains to Leo Fitz, "I'm not strong enough to live in a world that doesn't have you in it."[18]

In the end, as Todd May concludes, death "has the capacity, in a way no other aspect of us does, to absorb every other fact, to bring every other aspect of our lives under its sway."[19] Certainly, there are other important attributes—our capacity to create music or to write poems, for example. However, every facet of our mortal existence is conditioned by our finitude. This is a point that cannot be overemphasized in our understanding of being human in the MCU. The specter of death haunts the superhero genre, whether it is the death of humans and other-than-human beings, of societies and civilizations, or even of all life in the known universe. Indeed, on a regular basis the threat of Earth-bound or extraterrestrial extinction-level events are commonplace, as embodied by supervillains such as Thanos, Dormammu, and Kang the Conqueror.

Is critical reflection on meaningful life possible within the violent world of fictional superheroes? For many critical scholars, superheroes—notably Steve Rogers and Tony Stark—embody the apotheosis of humanism: able-bodied, cis-gendered, white men aligned with the military-industrial complex in the service of the nation state.[20] Consequently, in recent years, producers, directors, and screen writers have responded to the persistence of racial, gender, sexual, and ableist stereotypes embodied by superpowered characters, including many of the latest films and television series of the MCU. In the streaming series *Loki*, for example, the titular character is revealed as bisexual; likewise in *The Falcon and the Winter Soldier*, the racism encountered by Sam Wilson is confronted head-on; Matt Murdock (Daredevil) is visually impaired; and we learn (belatedly) that Hawkeye is hearing-impaired. Still, most efforts have been met with skepticism.[21] Samira Nadkarni, for example, cautions that while the MCU film and television franchise has worked toward greater diversity of nonwhite characters, issues of representation remain problematic, in that the MCU often continues to reinforce systems of a masculinist, white hegemonic power.[22] Writing specifically of the television series *Agents of S.H.I.E.L.D.*, Nadkarni explains that while the series "is arguably the best representation of racial diversity within the MCU," it falls short in that racialized "tropes, stereotypes, [and] historical allusions are often used in problematic ways to reinforce a social and cultural hegemony that continues to see the US. as largely white male led while simultaneously reinforcing American global dominance."[23]

I share these frustrations, but I also believe that the superhero genre, especially the MCU, provides an untapped resource to explore further the meaning of being human in an increasingly diverse posthuman universe. Sometimes, it is necessary to state the obvious: Not all characters in the superhero genre are superpowered. Standing alongside Captain America, Iron Man, Spider-Man, and the entire panoply of other-than-human and more-than-human characters are ordinary, nonenhanced humans: Nick Fury, Ellen Nadeer, Rosalind Price, Jimmy Woo, General Hale, Anton Ivanov. Whatever our expectations of a posthuman society might be, it is doubtful that everyone will embody extraordinary capabilities and it is important to consider how humanity will respond. As such, many of these critiques—while valid—prevent scholars from taking seriously other themes that are addressed in the superhero genre. Following Terence McSweeney, the superhero genre is worth studying, in part, because it reflects the times in which it appears and the social, political, and ideological factors that shape its narratives in meaningful ways.[24] As a form of speculative fiction, "superhero narratives can transcend the colorful and admittedly fantastical elements of their stories; this enables them to become vehicles for examining and deconstructing humanity's efforts to balance good and evil."[25] This includes the possibility of recognizing the will-to-live in both more-than-human and other-than-human beings.

That said, for many scholars and critics it remains nonsensical to take seriously the meaning of life and death in the superhero genre. Death, critics remark, is of no consequence in a genre where the dead are so easily brought back to life. Following the death of Tony Stark (Iron Man) in *Avengers: Endgame* (2019), for example, rumors began circulating on social media that the popular fictional character might reappear in future films. Could Iron Man once again cheat death and return from the grave? A growing chorus of fans speculated that Stark—prior to his demise—was able to digitize and transfer his consciousness to a computer—a procedure known as *mind-uploading*—and thereby return in another form. Although a common theme in speculative fiction, scientists are actively pursuing this "endgame" of human mortality.[26] The tech company Nectome, for example, is investing billions of dollars to preserve brains until solutions are found to upload their contents into computers.[27]

Within the fictional world of the MCU, the return of Tony Stark is not far-fetched. Indeed, the return of Iron Man requires nothing more than writing his character into the next film or streaming series. And this gets to a perennial critique of the superhero genre. If death provides meaning in life, what happens if (human) beings can "easily" return from the dead—not as monstrous, brain-eating zombies, but as the persons they once were? If death is reduced to nothing more than a momentary inconvenience, is it possible to say anything about meaningful life? During the climax of *Iron Man 3*, for example, Pepper Potts—Stark's business partner and love interest—seemingly falls to her death. However, as McSweeney writes, "This sense of precarity is a brief one." Potts does not die and, momentarily, is enhanced with superpowers. For McSweeney, this scene is characteristic of the superhero genre and calls to question whether life and death even matter in the MCU. "The idea that life in Hollywood cinema is fragile," McSweeney bemoans, "is just a momentary illusion; only in a Hollywood film could someone experience the trauma of the violent death of a loved one, just for it to be disavowed moments later and normalcy be reconstituted."[28]

McSweeney is not wrong. In the MCU, characters die, only to reappear later, sometimes with alarming regularity, in subsequent films or episodes; Steve Rogers, Bucky Barnes, Pepper Potts, Nick Fury, Leopold Fitz, Melinda May, Vision, and Gamora are just a few examples.[29] Certainly, some characters have died and (as of this writing) remain dead: Pietro Maximoff, Yondu, Natasha Romanoff, Maria Hill, May Parker, and Tony Stark. However, in the MCU, there is *always* the possibility of coming back from the dead, a feature that arguably has hampered the critical and sustained analysis of life and death in the superhero genre. Perhaps no character personifies the impermanence of death more than Phil Coulson. Coulson appears in *Iron Man* (2008), *Iron Man 2* (2010), and *Thor* (2011) as an agent of S.H.I.E.L.D. In these films, Coulson is a minor character, often providing comic relief with well-timed quips. In *The Avengers* (2012), however, the character of Coulson proves crucial, in that his death at the hands of Loki is used to motivate the superheroes to "assemble" and defeat an alien invasion. Coulson's death was pivotal to the narrative arc of the MCU. His resurrection would potentially diminish the meaningfulness of his sacrifice. However, within days of his cinematic death, fans flooded social media with a campaign (#CoulsonLives) to bring back Coulson. Subsequently, Coulson is resurrected in the Marvel television series *Agents of S.H.I.E.L.D.*, with his death and rebirth forming much of the narrative arc of the show. During the first season, for example, Coulson is brought back to life following an infusion of GH-325, a drug derived from an ancient Kree corpse. In season five, Coulson dies a "natural" death, after the effectiveness of the GH-325 drug was negated following Coulson's "possession" by the Ghost Rider in season four. Coulson—or at least the semblance of Coulson—reappears as an alien (known as Sarge) who is physically and genetically identical to Coulson. It is later revealed that the being known as Sarge was created in a cosmic fluke, a hybrid of Coulson's body and memories and that of another alien entity known as Pachakutiq. Pachakutig is subsequently killed and with it, the body of Coulson. This is not the "end" of Coulson, though. In the final season, Coulson reappears as a Life Model Decoy (an android), to help defeat an alien species, the Chronicoms, who threaten Earth. With a shrug of his shoulders, Coulson jests, "Dying. . . . it's kind of my super power."[30]

No doubt, some readers will remain skeptical. Still, I maintain that it is precisely the prospect of, if not actual, mortal violence, that gives the superhero genre its purchase to explore our posthuman future. In fact, it is precisely the mortal (and embodied) transformations—such as the recurrent deaths of Coulson—that open new vistas to contemplate our increasingly posthuman world. For each transformation of Coulson constitutes a

Death is no stranger to Phil Coulson (Clark Gregg) in the Marvel Cinematic Universe. His death in *The Avengers* (2012) is the first of many deaths for the character.

transformed process (and relation) of birth and death. That is to say, other characters respond differently to Coulson, contingent upon his "reappearance." Moreover, when a character *does* die—and *stays* dead—the emotional loss experienced by the survivors (and the audience) is heightened. When Rocket Raccoon watches his friends—Lylla the Otter, Teefs the Walrus, and Floor the Rabbit—get murdered in *Guardians of the Galaxy Vol. 3*, we experience his loss with him. Rocket knows, and we know, that there is no return, no salvation, no escape from death. As Coulson explains to Melinda May in *Agents of S.H.I.E.L.D.*, "The world is full of evil and lies and pain and death, and you can't hide from it—you can only face it. The question is. . . . how do you respond? Who do you become?"[31] Similarly, as the Mandarin ominously asks in *Iron Man 3*, "Do you want an empty life, or a meaningful death?"[32]

Lives, Todd May explains, are diachronic; they unfold over time.[33] Humans recognize life as having trajectories, of beginnings, of endings, and the finite time in between.[34] Death makes this recognition possible and, in doing so, conditions our capacity to find meaning in life. In other words, meaning in life is a process, not an endpoint, and it is the narrative character of our lives that can offer a source of meaningfulness. Essentially, what makes a life meaningful is precisely the coming together of those actions and inactions that make life possible. In addition, May explains, "If our lives are to be meaningful, then they must not only feel meaningful to us; they must also express a meaning that is not simply a matter of personal taste. They must have a worth that is grounded in something outside what you or I happen to enjoy or admire."[35] May expounds: "For a life to be meaningful, it also must possess a sense of subjective engagement.[36] In other words, finding meaning in life is relational. A meaningful life is a life shared with others, including those other-than-human beings. Accordingly, I suggest that humanity's self-awareness of death—either of one's own or of another—provides meaning in life and, consequently, mediates how humans interact with other-than-human beings, including both superheroes and supervillains but also the entire panoply of nonhumans that inhabit the MCU.

This sense of subjective engagement often arises when we confront life-and-death choices. Recall that when Kate Bishop in the streaming series *Hawkeye* asks Clint Barton about the best shot he ever made with his bow-and-arrow, the titular character responds simply, "The one I didn't take."[37] Here, Barton alludes to the time he was sent to kill Natasha Romanoff (Black Widow). Against orders, Barton did not take the shot, thus sparing Romanoff's life. In time, the two became close friends and, in a world beset with unimaginable violence, Barton and Romanoff would experience meaningfulness in each other's lives. Moreover, Bishop herself aspires to become a superhero following the traumatic death of her farther. As a young child, Bishop and her parents lived in New York when extraterrestrial beings—the Chitauri—attempted to invade Earth. Bishop watched in awe as Hawkeye, along with the other Avengers, risked their own lives to protect humanity. Tragically, Bishop's father died in the battle. His death, though, altered the life-course of Bishop and gave meaning both to her life and of the memory of her father. And recall also that Ellen Nadeer experienced traumatic loss in the Battle of New York. For Nadeer, however, death fueled a hatred of the alien Other, a hatred that extended to Inhumans.

Schweitzer's notion of the will-to-live resonates strongly with the superhero genre insofar as the genre explicitly confronts the story arc of characters, tracing their individual "origin" stories to, perhaps, their own demise. The life-and-death choices we make, similar to the choices made by Bishop and Barton, are constitutive of the narrative value of finding meaning both in life and in death. As such, within the superhero genre, origin stories do more than simply describe how a character gained their super powers; rather, these stories narrate lives made meaningful through moral action and inaction.

Morality, May explains, "Concerns our living. It tells us ways in which we should or should not go about it and draws lines between what is permissible and what is not. It concerns our acts or the trajectory of our lives. It tells us what we ought to do or how we ought to develop ourselves. Narrative values, while they are not matters of *should* or *ought* in the same way moral values are, are focused on living in ways that will make life more valuable in a particular

Life Model Decoy Melinda May (Ming-Na Wen) laments that she is afraid to die in *Agents of S.H.I.E.L.D.* (2017).

way—the way of meaningfulness."[38] Following Schweitzer, a person is truly ethical only when they obey the compulsion to help all life which they are able to assist, and refrains from harming anything that lives. Superheroes, including those in the MCU, largely accept their responsibility toward life. That is to say, they do not expressly attempt to do harm. Certainly, innocent lives are lost in the inevitable clash between superheroes and supervillains and, increasingly, Avengers—notably Tony Stark and Bruce Banner—express guilt and grief following their actions. Ordinary humans, however, are a decidedly mixed bag. Those who hold extremist views of human exceptionalism, for example, fail to recognize the will-to-live in more-than-human and other-than-human beings, deciding instead to incarcerate or exterminate the nonhumans as an existential threat.

The will-to-live among nonhuman beings is expressed poignantly by the Life Model Decoy Melinda May who, belatedly, learns that "she" is not human, but instead is an android. She asks Coulson (still very much a human), "Are you afraid to die? Because I am." The android continues: "I know I'm not real. I'm all phantom limbs. But that doesn't make the pain less real."[39] And yet, she is real. Although composed of electrical circuits and software instead of flesh and bone, her body will eventually degrade. Objects such as computers and washing machines have lifespans, a finite period until their constituent parts wear out. Markedly though, in science fiction the death of a robot often differs from that of a human. According to Lyons, this is because the robot's death "does not carry the same weight, significance, or consideration as a biological death" experienced by humans. Lyons explains, however, that "this is part of what makes the robot death both so poignant and so tragic; the demise occurs but without being considered real or authentic, further plaguing the nonhuman character for whom death remains an occurrence all the same."[40] Crucially, the sentient robot does not conceive of itself as artificial; as we see with LMD Melinda May, she grasped her own unique authenticity, a reality denied her by human counterparts. As philosophers and other scientists suggest, we may not ever know fully if other sentient beings are aware of their own mortality.[41] But if these beings are sentient, there stands a good possibility that, like the LMD Melinda May, they are aware of their finitude.

In Marvel's *Agents of S.H.I.E.L.D.*, Simmons contemplates her own mortality. She says to Fitz, "I like to think about the first law of thermodynamics, that no energy in the universe is created and . . . none is destroyed." For Simmons, this simple statement has tremendous implications in giving meaning to life—and death. She continues: "That means that every bit of energy inside us, every particle, will go on to be a part of something else. Maybe live as a dragonfish, a microbe, maybe burn in a supernova ten billion years from now. And every part of us now was once a part of some other thing—a moon, a storm cloud, a mammoth." Fitz interjects: "A monkey." "A monkey," Simmons agrees, "Thousands and thousands of other beautiful things that were just as terrified to die as we are."[42]

As humans and society become increasingly technological and virtual, how do we—or can we—define what constitutes the human in relation to other forms of being, including beings organic or inorganic, terrestrial or extraterrestrial? Perhaps humans are truly exceptional in the knowledge of our finitude. But in a way, it doesn't matter. What's important is that humans *are* aware of *our* future death and the future death of others, and this knowledge mediates our mortal existence. Simply stated, the superhero genre—but especially the MCU—provides a remarkable opportunity to ponder humanity's ethical positions and moral obligations toward more-than-human and other-than-human beings, whether these be mutants, androids, or aliens, and to recognize their will-to-live. The MCU, ultimately, offers an opportunity to critically evaluate the meaning of being human in the ghastly drama of superhero fiction.

NOTES

PREFACE

1. S.H.I.E.L.D. is the somewhat contrived acronym for a secretive government agency in the Marvel Cinematic Universe: the Strategic Homeland Intervention Enforcement and Logistics Division. Created by Stan Lee and Jack Kirby in 1965, the acronym originally stood for the Supreme Headquarters, International Espionage and Law-Enforcement Divison. It was rebranded in 1991 as the Strategic Hazards Intervention Espionage Logistics Directorate. It assumed its current incarnation in 2008 with the cinematic release of *Iron Man*. See Darren Franich, "SHIELD: 10 Important Facts about Marvel's Superspy Organization," *Entertainment Weekly*, September 24, 2013, ew.com/article/2013/09/24/shield-agents-marvel/ (accessed May 13, 2023).

2. Russell Blackford, *Science Fiction and the Moral Imagination: Visions, Minds, Ethics* (Cham, Switzerland: Springer 2017), 130.

3. Juli L. Gittinger, *Personhood in Science Fiction: Religious and Philosophical Considerations* (Cham, Switzerland: Palgrave Macmillan, 2019), 180.

4. Robert Pepperell, *The Post-Human Condition* (Bristol, UK: Intellect Books, 2003); Rosi Braidotti, *The Posthuman* (Cambridge, MA: Polity Press, 2013); Stefan Herbrechter, *Posthumanism: A Critical Analysis* (New York: Bloomsbury, 2013); Pramod K. Nayar, *Posthumanism* (Malden, MA: Polity Press, 2014); Rosi Braidotti, *Posthuman Knowledge* (Malden, MA: Polity Press, 2019); Rosi Braidotti, "'We' Are in *This* Together, But We Are Not One and the Same," *Bioethical Inquiry* 17, no. 4 (2020): 465–69; and Francesca Ferrando, *Philosophical Posthumanism* (New York: Bloomsbury, 2020).

5. Pepperell, *The Post-Human Condition*, vi.

6. Pepperell, *The Post-Human Condition*, 1.

7. Adam Weitzenfeld and Melanie Joy, "An Overview of Anthropocentrism, Humanism, and Speciesism in Critical Animal Theory," *Counterpoints* 448 (2014): 3–27.

8. David Abram, *The Spell of the Sensuous: Perception and Language in a More-Than-Human World* (New York: Pantheon, 1996).

9. Nayar, *Posthumanism*, 4.

10. James Bridle, *Ways of Being: Animals, Plants, Machines: The Search for a Planetary Intelligence* (New York: Picador, 2022), 17.

11. For more on "anti-humanism," see for example Douglas V. Porpora, "Dehumanization in Theory: Anti-Humanism, Non-Humanism, Post-Humanism, and Trans-Humanism," *Journal of Critical Realism* 16, no. 4 (2017): 353–67.

12. This is the theme of many speculative films, including *Elysium* (2013) and *Don't Look Up* (2021).

13. Braidotti, *Posthuman Knowledge*, 11.

14. Gavin J. Andrews and Andrea Rishworth, "New Theoretical Terrains in Geographies of Wellbeing: Key Questions of the Posthumanist Turn," *Wellbeing, Space and Society* 4 (2023): 100130.

15. Esther De Dauw, *Hot Pants and Spandex Suits: Gender Representation in American Superhero Comic Books* (New Brunswick, NJ: Rutgers University Press, 2021). See also Jason Dittmer, "Captain America in the News: Changing Mediascapes and the Appropriation of a Superhero," *Journal of Graphic Novels and Comics* 3, no. 2 (2012): 143–57.

16. Samira Shirish Nadkarni, "'To Be the Shield': American Imperialism and Explosive Identity Politics in *Agents of S.H.I.E.L.D.*," in *Assembling the Marvel Cinematic Universe: Essays on the Social, Cultural and Geopolitical Domains*, ed. Julian C. Chambliss, William L. Svitavsky, and Daniel Fandino (Jefferson, NC: McFarland & Co., 2018), 219.

17. How successful these efforts have been remain subject to debate among both fans and scholars alike. See for example Casey M. Ratto, "Not Superhero Accessible: The Temporal Stickiness of Disability in Superhero Comics," *Disability Studies Quarterly* 37, no. 2 (2017): n.p.; Robyn Joffe, "Holding Out for a Hero (ine): An Examination of the Presentation and Treatment of Female Superheroes in Marvel Movies," *Panic at the Discourse: An Interdisciplinary Journal* 1, no. 1 (2019): 5–19; Lewis Call, "'Alien Commies from the Future!': Diversity, Equity, and Inclusion in Season Seven of *Agents of S.H.I.E.L.D.*," *Slayage* 19, nos. 1–2 (2021): 143–84; Viani Alifa Almas, "Reconstructing The Concept of Superhero in Ms. Marvel TV Series: First Female Moslem Superhero's Journey," *LITERA KULTURA: Journal of Literary and Cultural Studies* 10, no. 3 (2022): 12–17; Liam Drislane, "The Pretense of Prosthesis: The Prosthecized Superhero in the Marvel Cinematic Universe," *Science Fiction Film and Television* 15, no. 2 (2022): 169–91; and Stephanie Green, "Playing at Being a Superhero: Trish Walker in Jessica Jones," *Imagining the Impossible: International Journal for the Fantastic in Contemporary Media* 1, no. 1 (2022): 1–16.

18. Jason Dittmer, "On Captain America and 'Doing' Popular Culture in the Social Sciences," *E-International Relations*, May 5, 2015, e-ir.info/2015/05/05/on-captain-america-and-doing-popular-culture-in-the-social-sciences/ (accessed November 1, 2022). See also Mitch Murray, "The Work of Art in the Age of the Superhero," *Science Fiction Film and Television* 10, no. 1 (2017): 27–51.

19. Quoted in Mayra Garcia, "10 Criticisms of the MCU from Celebrities (& Why Each is Off the Mark)," *CBR.Com*, December 13, 2021, www.cbr.com/criticisms-of-marvel-from-celebrities/ (accessed November 1, 2022).

20. Martin Scorsese, "Martin Scorsese: I Said Marvel Movies Aren't Cinema. Let Me Explain," *New York Times*, November 4, 2019, www.nytimes.com/2019/11/04/opinion/martin-scorsese-marvel.html (accessed November 1, 2022).

21. Catherine Shoard, "Francis Ford Coppola: Scorsese was Being Kind—Marvel Movies are Despicable," *The Guardian*, Oct. 21, 2019. www.theguardian.com/film/2019/oct/21/francis-ford-coppola-scorsese-was-being-kind-marvel-movies-are-despicable (accessed

November 1, 2022). See also Julian Lawrence and Neil Archer, "Martin Scorsese Says Superhero Movies are 'Not Cinema': Two Experts Debate," *The Conversation*, October 24, 2019, theconversation.com/martin-scorsese-says-superhero-movies-are-not-cinema-two-experts-debate-125771 (accessed November 1, 2022).

22. What is or isn't canon in the MCU is subject to intense and passionate debate among its aficionados. See for example Thomas Bacon, "Why Marvel's Disney+ Shows are Rewriting MCU History," *Screenrant*, March 30, 2021, screenrant.com/marvel-disney-plus-mcu-history-change-how-reasons/; Charles Cameron, "All 12 MCU TV Shows That Aren't Canon (But Should Be)," *Screenrant,* October 3, 2021, screenrant.com/marvel-tv-shows-mcu-canon-no-netflix-why/; Savannah Sanders, "Marvel Studios Author Disputes *Agents of S.H.I.E.L.D.'s* Place in MCU Canon," *The Direct*, October 21, 2021, thedirect.com/article/marvel-agents-of-shield-mcu-canon-question; Jennifer McDonough, "Hawkeye Directors Dodge Question about Daredevil's MCU Canoncity," *The Direct*, December 17, 2021, thedirect.com/article/hawkeye-daredevil-mcu-canon-question; and Benny Stein, "Marvel Studios is Abandoning MCU Canon with New Disney+ Show," *The Direct*, June 5, 2022, thedirect.com/article/marvel-studios-mcu-canon-disney-show.

23. Daniel W. Drezner, "What the Marvel Cinematic Universe Can Teach Us About Geopolitics," *foreignpolicy.com*, February 19, 2023, foreignpolicy.com/2023/02/19/marvel-cinematic-universe-politics-international-relations-superheroes-book-review/ (accessed December 18, 2023).

24. Vera Veldhuizen, "Classifying Monsters," in *Interrogating Boundaries of the Nonhuman: Literature, Climate Change, and Environmental Crises*, ed. Matthias Stephan and Sune Borkfelt (Lanham, MD: Lexington Press, 2022), 166. See also Sherryl Vint, *Animal Alterity: Science Fiction and the Question of the Animal* (Liverpool, UK: Liverpool University Press, 2010), 4.

25. Catherine Marshall, "Refugees and Other Aliens," *Eureka Street* 19, no. 16 (2009): 39.

26. Nayar, *Posthumanism*, 8.

27. Susan Wolf, *Meaning in Life and Why It Matters* (Princeton, NJ: Princeton University Press, 2010); Thaddeus Metz, *Meaning in Life: An Analytic Study* (Oxford, UK: Oxford University Press, 2013); Todd May, *A Significant Life: Human Meaning in a Silent Universe* (Chicago: University of Chicago Press, 2015); Martin Hägglund, *This Life: Why Mortality Makes Us Free* (London: Profile Books, 2019).

28. Martin Schönfeld, "Who or What has Moral Standing?" *American Philosophical Quarterly* 29, no. 4 (1992): 353.

29. Roderick Frazier Nash, *The Rights of Nature: A History of Environmental Ethics* (Madison: University of Wisconsin Press, 1989); J. Claude Evans, *With Respect for Nature: Living as Part of the Natural World* (Albany: State University of New York Press, 2005); Robin Attfield, *Environmental Ethics: A Very Short Introduction* (New York: Oxford University Press, 2018); Gregory Bassham, *Environmental Ethics: The Central Issues* (Indianapolis: Hackett Publishing, 2020).

30. Terence McSweeney, *The Contemporary Superhero Film: Projections of Power and Identity* (New York: Wallflower, 2020), 25.

31. I'm sensitive to the counterargument, that Stark is irredeemably bad, despite his apparent maturation in later films. Captain America, also, is for many viewers the embodiment of a white, Christian military ethos that saturates American society.

NOTES

32. Nash, *The Rights of Nature*, 7.

33. George Seaver, *Albert Schweitzer: The Man and His Mind* (London: A&C Black, 1947); Henry Clark, *The Ethical Mysticism of Albert Schweitzer* (Boston: Beacon Press, 1962); James Brabazon, *Albert Schweitzer: A Biography* (Syracuse, NY: Syracuse University Press, 2000); Ara Barsam, *Reverence for Life: Albert Schweitzer's Great Contribution to Ethical Thought* (Oxford: Oxford University Press, 2008); Mike W. Martin, *Albert Schweitzer's Reverence for Life: Ethical Idealism and Self-Realization* (Aldershot, UK: Ashgate, 2012); Nils Ole Oermann, *Albert Schweitzer: A Biography* (Oxford: Oxford University Press, 2016); and James Carleton Paget and Michael J. Thate, eds., *Albert Schweitzer in Thought and Action: A Life in Parts* (Syracuse, NY: Syracuse University Press, 2016).

34. Albert Schweitzer, *The Philosophy of Civilization*, trans. C. T. Campion (Amherst, NY: Prometheus Books, 1987), 309.

35. Barsam, *Reverence for Life*, x.

36. John R. Everett, "Albert Schweitzer and Philosophy," *Social Research* 33, no. 4 (1966): 516.

37. Barsam, *Reverence for Life*, 9.

38. Martin, *Albert Schweitzer's Reference for Life*, 3.

39. Schweitzer, *The Philosophy of Civilization*, 332.

40. Joanna Robinson, Dave Gonzales, and Gavin Edwards, *MCU: The Reign of Marvel Studios* (New York: Liveright Publishing, 2024), 4.

41. Martin Flanagan, Mike McKenny, and Amy Livingstone, *The Marvel Studios Phenomenon: Inside a Transmedia Universe* (New York: Bloomsbury, 2016); Jeffrey A. Brown, *The Modern Superhero in Film and Television: Popular Genre and American Culture* (New York: Routledge, 2017); Julian C. Chambliss, William L. Svitavsky, and Daniel Fandino, eds., *Assembling the Marvel Cinematic Universe: Essays on the Social, Cultural and Geopolitical Domains* (Jefferson, NC: McFarland & Company, 2018); and Terence McSweeney, *Avengers Assemble!: Critical Perspectives on the Marvel Cinematic Universe* (New York: Columbia University Press, 2018).

42. "Box Office History for Marvel Cinematic Universe Movies, the-numbers.com/movies/franchise/Marvel-Cinematic-Universe#tab=summary (accessed March 27, 2022).

43. Boxofficemojo.com, boxofficemojo.com/genre/sg2900226305/?ref_=bo_gs_table_8 (accessed March 27, 2022).

44. Boxofficemojo.com, https://www.boxofficemojo.com/chart/top_lifetime_gross/?ref_=bo_cso_ac (accessed March 27, 2022).

45. *Avengers* (2012).

CHAPTER ONE: HUMANS, ASSEMBLE!

1. *Avengers: Endgame* (2019).

2. *Avengers: Endgame* (2019).

3. Joseph Zornado and Sara Reilly, *The Cinematic Superhero as Social Practice* (Cham, Switzerland: Palgrave Macmillan, 2021), 29.

NOTES

4. Terence McSweeney, *Avengers Assemble! Critical Perspectives on the Marvel Cinematic Universe* (New York: Columbia University Press, 2018), 163–64.

5. McSweeney, *Avengers Assemble*, 217.

6. Reed Tucker, *Slugfest: Inside the Epic 50-Year Battle Between Marvel and DC* (London: Sphere, 2017), 222.

7. Haoyang Wang and Christina Zhang, "Marvel Cinematic Universe Villains and Social Anxieties," in *The Politics of the Marvel Cinematic Universe*, ed. Nicholas Carnes and Lilly J. Goren (Lawrence: University Press of Kansas, 2023).

8. In fairness to Vision, he states that "there may be no way to make you trust me"— even as he lifts Thor's hammer, thus demonstrating his worthiness.

9. See for example Jason Dittmer, "American Exceptionalism, Visual Effects, and the Post-9/11 Cinematic Superhero Boom," *Environment and Planning D: Society and Space* 29, no. 1 (2011): 114–30; Ashley S. Robinson, "We Are Iron Man: Tony Stark, Iron Man, and American Identity in the Marvel Cinematic Universe's Phase One Films," *The Journal of Popular Culture* 51, no. 4 (2018): 824–44.

10. Robinson, "We Are Iron Man," 825.

11. Zornado and Reilly, *The Cinematic Superhero*, 97.

12. Zornado and Reilly, *The Cinematic Superhero*, 103.

13. Nicholas Carnes and Lilly J. Green, "An Introduction to the Politics of the Marvel Cinematic Universe," in *The Politics of the Marvel Cinematic Universe*, ed. Nicholas Carnes and Lilly J. Goren (Lawrence: University Press of Kansas, 2023), 9.

14. Science fiction, and especially the superhero genre, opens the door to contemplate immaterial life, that is, a living "thing" that has neither mass nor volume. For our present purposes, however, we'll limit our discussion to the material universe.

15. Peter Shaver, *Cosmic Heritage: Evolution from the Big Bang to Conscious Life* (London: Springer, 2011), 1. See also Sun Kwok, *Stardust: The Cosmic Seeds of Life* (Berlin: Springer, 2013).

16. Shaver, *Cosmic Heritage*, 55.

17. Shaver, *Cosmic Heritage*, 61.

18. David Baker, *The Shortest History of Our Universe: The Unlikely Journey from the Big Bang to Us* (New York: The Experiment, 2023), 5.

19. Shaver, *Cosmic Heritage*, 26–32.

20. Shaver, *Cosmic Heritage*, 79.

21. There is a vast literature on the constitution of life. For scientific accounts, see Erwin Schrödinger, *What is Life? The Physical Aspect of the Living Cell and Mind* (Cambridge: Cambridge University Press, 1944); Robert Hazen, "What is Life?" *New Scientist* 192 (2006): 46–51; Bill Mesler and H. James Cleaves II, *A Brief History of Creation: Science and the Search for the Origin of Life* (New York: W. W. Norton & Company, 2016); Paul Nurse, *What is Life? Five Great ideas in Biology* (New York: W. W. Norton & Company, 2021). For more philosophical interpretations, see John P. Lizza, *Persons, Humanity, and the Definition of Death* (Baltimore: Johns Hopkins University Press, 2006); Steven Luper, *The Philosophy of Death* (Cambridge: Cambridge University Press, 2009); Bernard M. Schumacher, *Death and Mortality in Contemporary Philosophy*, trans. Michael J. Miller (Cambridge: Cambridge

University Press, 2011). Françoise Dastur, *How Are We to Confront Death? An Introduction to Philosophy*, trans. Robert Vallier (New York: Fordham University Press, 2012).

22. Christopher F. Chyba and Kevin P. Hand, "Astrobiology: The Study of the Living Universe," *Annual Review of Astronomy and Astrophysics* 43 (2005): 39.

23. Nurse, *What is Life*, 135, 143.

24. Mesler and Cleaves, *A Brief History of Creation*, 164. The other two percent is made up of trace elements, notably sulfur and phosphorus.

25. Mark Lupisella, "Astrobiology and Cosmocentrism," *Bioastronomy News* 10, no. 1 (1998): 8.

26. Jamie Metzl, *Hacking Darwin: Genetic Engineering and the Future of Humanity* (Naperville, IL: Sourcebooks, 2019), 7.

27. Nurse, *What is Life*, 44.

28. Shaver, *Cosmic Heritage*, 117.

29. Metzl, *Hacking Darwin*, 7.

30. Nurse, *What is Life*, 41.

31. Amy Chambers, "You Gotta Make Way for the Homo Superior: Mutation, Evolution, and Super Powers on Screen," May 30, 2018, amycchambers.com/2018/05/30/mutation-evolution-and-super-powers-on-screen/ (accessed May 11, 2023).

32. Rosi Braidotti, *Posthuman Knowledge* (Malden, MA: Polity Press, 2019), 1.

33. George Lakoff, "Introduction," in *Louder than Words: The New Science of How the Mind Makes Meaning*, ed. Benjamin K. Bergen (New York: Basic Books, 2012), ix–xi.

34. Elana Gomel, "Posthuman Rights: The Ethics of Alien Encounter," in *Unveiling the Posthuman*, ed. Artur Matos Alves (Oxford: Inter-Disciplinary Press, 2012), 12.

35. Stephen J. Gould, "The Evolution of Life on the Earth," *Scientific American* 271, no. 4 (1994): 86.

36. Gould, "The Evolution of Life," 86.

37. Gould, "The Evolution of Life," 87.

38. Julie Galway-Witham and Chris Stringer, "How Did *Homo sapiens* Evolve?" *Science* 360, no. 6395 (2018): 1296–98.

39. Pregrag B. Slijepčevič, *Biocivilizations: A New Look at the Science of Life* (White River Junction, VT: Chelsea Green, 2023), 2.

40. Patrick R. Frierson, *What is the Human Being?* (New York: Routledge, 2013); Marek Tamm and Zoltán Boldizsár Simon, "Historical Thinking and the Human: Introduction," *Journal of the Philosophy of History* 14, no. 3 (2020): 285–309.

41. See for example James L. Mays, "What is a Human Being? Reflections on Psalm 8," *Theology Today* 50, no. 4 (1994): 511–20.

42. Lynn White Jr., "The Historical Roots of Our Ecological Crisis," *Science* 155, no. 3767 (1967): 1206.

43. Sue Donaldson and Will Kymlicka, "The Moral Ark," *Queen's Quarterly* 114, no. 2 (2007): 192. See also Daniel J. Wilson, "Lovejoy's the Great Chain of Being After Fifty Years," *Journal of the History of Ideas* 48, no. 2 (1987): 187–206 and Sean Nee, "The Great Chain of Being," *Nature* 435, no. 7041 (2005): 429.

44. Paul Robbins, John Hintz, and Sarah A. Moore, *Environment and Society: A Critical Introduction* (New York: Wiley Blackwell, 2014), 68.

45. Roderick F. Nash, *The Rights of Nature: A History of Environmental Ethics* (Madison: University of Wisconsin Press, 1989), 17.

46. Gregory Bassham, *Environmental Ethics: The Central Issues* (Indianapolis: Hackett Publishing, 2020), 26.

47. Robbins, Hintz, and Moore, *Environment and Society*, 69.

48. Bassham, *Environmental Ethics*, 36.

49. J. Baird Callicott, "Traditional American Indian and Western European Attitudes Toward Nature: An Overview," *Environmental Ethics* 4 (1982): 306. See also Soren C. Larsen and Jay T. Johnson, "In Between Worlds: Place, Experience, and Research in Indigenous Geography," *Journal of Cultural Geography* 29, no. 1 (2012): 1–13; Rani-Henrik Andersson, Boyd Cothran and Saara Kekki, eds., *Bridging Cultural Concepts of Nature: Indigenous People and Protected Spaces of Nature* (Helsinki: Helsinki University Press, 2021); Nancy J. Turner and Andrea J. Reid, "'When the Wild Roses Bloom': Indigenous Knowledge and Environmental Change in Northwestern North America," *GeoHealth* 6, no. 11 (2022): 1–21; and Niiyokamigaabaw Deondre Smiles, "Reflections on the (Continued and Future) Importance of Indigenous Geographies," *Dialogues in Human Geography* 14, no. 2 (2024): 217–20.

50. Donaldson and Kymlicka, "The Moral Ark," 192.

51. Daniel Rueda Garrido, "Deaths of the Subject and Negated Subjectivity in the Era of Neoliberal Capitalism," *tripleC* 17, no. 1 (2019): 159–84.

52. Affrica Taylor, "Romancing or Re-Configuring Nature in the Anthropocene? Towards Common Worlding Pedagogies," in *Reimagining Sustainability in Precarious Times*, ed. Karen Malone, Son Truong, and Tonia Gray (Singapore: Springer Nature, 2017), 62. (Emphasis in original.)

53. Pramod Nayar, *Posthumanism* (Malden, MA: Polity Press, 2014), 4.

54. Zakiyyah Iman Jackson, "Outer Worlds: The Persistence of Race in Movement 'Beyond the Human,'" *GLQ: A Journal of Lesbian and Gay Studies* 21, no. 2 (2015): 215.

55. Rosi Braidotti, "The Posthuman Predicament," in *The Scientific Imaginary in Visual Culture*, edited by Anneke Smelik (Goettingen: V&R Press, 2010), 71; see also Patricia Hill Collins, *Black Feminist Thought: Knowledge, Consciousness and the Politics of Empowerment* (New York: Routledge, 1991); Paul Gilroy, *Against Race: Imagining Political Culture Beyond the Color Line* (Cambridge, MA: Harvard University Press, 2000); Sylvia Wyner, "Unsettling the Coloniality of Being/Power/Truth/Freedom: Toward the Human, After Man, Its Overrepresentation—An Argument," *CR: The New Centennial Review* 3, no. 3 (2003): 257–337; Zakiyyah Iman Jackson, "Animal: New Directions in the Theorization of Race and Posthumanism," *Feminist Studies* 39, no. 3 (2013): 669–85.

56. Rosi Braidotti, "'We' Are in *This* Together, But We Are Not One and the Same," *Bioethical Inquiry* 17, no. 4 (2020): 467.

57. Patricia MacCormack, "Queer Posthumanism: Cyborgs, Animals, Monsters, Perverts," in *The Ashgate Research Companian to Queer Theory*, ed. Noreen Giffney and Michael O'Rourke (New York: Routledge, 2009), 111.

58. Braidotti, "'We' May Be," 29.

59. Sylvia Wynter, "No Humans Involved: An Open Letter to my Colleagues," *Forum N.H.I.* 1, no. 1 (1994): 42–73; Alida Weidensee, "No Humans Involved: The Dehumanization

of Missing and Murdered Indigenous Women and Girls," *Minnesota Journal of Law and Inequality* (2021), available at lawandinequality.org/2021/11/29/no-humans-involved-the-dehumanization-of-missing-and-murdered-indigenous-women-and-girls/ (accessed October 29, 2022).

60. Gomel, "Posthuman Rights," 12.

61. Robert Pepperell, *The Post-Human Condition* (Bristol, UK: Intellect Books, 2003), 177.

62. Charles S. Cockell, "Microbial Rights?" *EMBO Reports* 12, no. 3 (2011): 181.

63. Bernard Dixon, "Smallpox-Imminent Extinction and an Unresolved Dilemma," *New Scientist* 69, no. 989 (1976): 430–32.

64. Cockell, "Microbial Rights," 181.

65. Moral standing, as introduced in the preface, has never been universal in human history and ethical rights have always been severely circumscribed, with only select members of humanity accord full membership. In other words, the conceit of being human has always been exclusionary, marked by racism, sexism, classism, ageism, ableism, homophobia, and other prejudiced beliefs purportedly backed by science and enforced by the law.

66. Nash, *The Rights of Nature*, 18.

67. Elana Gomel, *Science Fiction, Alien Encounters, and the Ethics of Posthumanism: Beyond the Golden Rule* (New York: Palgrave Macmillan, 2014), 23. See also Christoper D. Stone, "Should Trees have Standing?—Toward Legal Rights for Natural Objects," *Southern California Law Review* 45 (1972): 450–501; Charlie R. Crimston, Paul G. Bain, Matthew J. Hornsey, and Brock Bastian, "Moral Expansiveness: Examining Variability in the Extension of the Moral World," *Journal of Personality and Social Psychology* 111, no. 4 (2016): 636–53; Charlie R. Crimston, Matthew J. Hornsey, Paul G. Bain, and Brock Bastian, "Toward a Psychology of Moral Expansiveness," *Current Directions in Psychological Science* 27, no. 1 (2018): 14–19; and Jacy Reese Anthis and Eze Paez, "Moral Circle Expansion: A Promising Strategy to Impact the Far Future," *Futures* 130 (2021): 102–56.

68. Nash, *The Rights of Nature*, 20–22.

69. Matthieu Ricard, *A Plea for the Animals: The Moral, Philosophical, and Evolutionary Imperative to Treat All Beings with Compassion* (Boulder, CO: Shambhala Publications, 2016), 30.

70. Schönfeld, "Who or What has Moral Standing," 356.

71. Peter Singer, *Animal Liberation Now: The Definitive Classic Renewed* (New York: Harper Perennial, 2015), 7. See also Peter Singer, *Animal Liberation: A New Ethics for our Treatment of Animals* (New York: New York Review, 1975).

72. Singer, *Animal Liberation Now*, 4–5.

73. Gary L. Francione and Anna E. Charlton, "The Case Against Pets," *Aeon*, September 8, 2016, aeon.co/essays/why-keeping-a-pet-is-fundamentally-unethical; Anna E. Charlton and Gary L. Francione, "A 'Humanely' Killed Animal is Still Killed—and That's Wrong," *Aeon*, September 8, 2017, aeon.co/ideas/a-humanely-killed-animal-is-still-killed-and-thats-wrong; Jonathan Birch, "Crabs and Lobsters Deserve Protection from Being Cooked Alive," *Aeon*, November 3, 2017, aeon.co/ideas/crabs-and-lobsters-deserve-protection-from-being-cooked-alive; Lori Marino, "They Are Prisoners," *Aeon*, February 2, 2021, aeon.co/essays/concrete-tanks-are-torture-for-social-intelligent-killer-whales; Jeff Sebo and Jason Schukraft, "Don't Farm Bugs," *Aeon*, July 27, 2021, aeon.co/essays/on-the-torment-of-insect-minds-and-our-moral-duty-not-to-farm-them;

Gary L. Francione, "We Must Not Own Animals," *Aeon*, March 1, 2022, aeon.co/essays/why-morality-requires-veganism-the-case-against-owning-animals; and Garet Lahvis, "Freefall into Darkness," *Aeon*, June 2, 2022, aeon.co/essays/what-do-caged-animals-really-tell-us-about-our-mental-lives (all accessed November 11, 2022).

74. Joel Feinberg, *Rights, Justice, and the Bounds of Liberty: Essays in Social Philosophy* (Princeton, NJ: Princeton University Press, 1980), 177.

75. Peter Singer, *Practical Ethics*, (New York: Cambridge University Press, 1993), 57.

76. Bassham, *Environmental Ethics*, 38.

77. John Rodman, "The Liberation of Nature?" *Inquiry* 20, no. 1–4 (1977): 83–131; Eric Katz, "The Liberation of Humanity and Nature," *Environmental Values* 11, no. 4 (2002): 397–405; Raffaele Rodogno, "Sentientism, Wellbeing, and Environmentalism," *Journal of Applied Philosophy* 27, no. 1 (2010): 84–99; and Andrew Y. Lee, "Speciesism and Sentientism," *Journal of Consciousness Studies* 29, nos. 3–4 (2022): 205–28.

78. Bassham, *Environmental Ethics*, 3. See also Stuart Russell, *Human Compatible: Artificial Intelligence and the Problem of Control* (New York: Penguin Books, 2020).

79. Ricard, *A Plea for the Animals*, 95.

80. Schönfeld, "Who or What has Moral Standing," 361.

81. James Bridle, *Ways of Being: Animals, Plants, Machines: The Search for a Planetary Intelligence* (New York: Picador, 2022), 10–11.

82. Schönfeld, "Who or What has Moral Standing," 361.

83. Bassham, *Environmental Ethics*, 3.

84. See for example John P. Lizza, *Persons, Humanity, and the Definition of Death* (Baltimore: Johns Hopkins University Press, 2006); Steven Luper, *The Philosophy of Death* (Cambridge: Cambridge University Press, 2009); Bernard M. Schumacher, *Death and Mortality in Contemporary Philosophy*, trans. by Michael J. Miller (Cambridge: Cambridge University Press, 2011); and Françoise Dastur, *How Are We to Confront Death? An Introduction to Philosophy*, trans. by Robert Vallier (New York: Fordham University Press, 2012).

85. Paul Walker and Terence Lovat, "Concepts of Personhood and Autonomy as They Apply to End-of-Life Decisions in Intensive Care," *Medical Health Care and Philosophy* 18 (2015): 310.

86. Luper, *The Philosophy of Death*, 14.

87. James L. Bernat, "The Biophilosophical Basis of Whole-Brain Death," *Social Philosophy and Policy* 19, no. 2 (2002): 333.

88. See for example Dave Fagundes, "What We Talk About When We Talk About Persons: The Language of a Legal Fiction," *Harvard Law Review* 114, no. 6 (2001): 1745–68; Linda MacDonald Glenn, "What is a Person?" in *Posthumanism: The Future of Homo Sapiens*, ed. Michael Bess and Diana Walsh Pasulka (Farmington Hills, MI: Macmillan, 2018), 229–46; and Tyler L. Jaynes, "On Human Genome Manipulation and *Homo technicus*: The Legal Treatment of Non-Natural Human Subjects," *AI and Ethics* 1 (2021): 331–45.

89. Jamie Morgan, "Species Being in the Twenty-First Century," *Review of Political Economy* 30, no. 3 (2018): 387.

90. See for example Gianmarco Veruggio and Fiorella Operto, "Roboethics: A Bottom-Up Interdisciplinary Discourse in the Field of Applied Ethics in Robotics," *International Review of Information Ethics* 6, no. 12 (2006): 2–8; Michael Anderson and Susan Leigh

Anderson, "Machine Ethics: Creating an Ethical Intelligent Agent," *AI Magazine* 28, no. 4 (2007): 15–26; Michael Anderson and Susan Leigh Anderson, eds., *Machine Ethics* (Cambridge: Cambridge University Press, 2011); David J. Gunkle, *The Machine Question: Critical Perspectives on AI, Robots, and Ethics* (Cambridge, MA: The MIT Press, 2017); Mark Coeckelbergh, *AI Ethics* (Cambridge, MA: The MIT Press, 2020); Rajakishore Nath and Vineet Sahu, "The Problem of Machine Ethics in Artifiical Intelligence," *AI & Society* 35 (2020): 103–11; Sven Nyholm, *Humans and Robots: Ethics, Agency, and Anthropomorphism* (New York: Rowman & Littlefield, 2020); and Mark Coeckelbergh, *Robot Ethics* (Cambridge, MA: The MIT Press, 2022).

91. *Daredevil*, Season 2, Episode 3, "New York's Finest."

92. *Daredevil*, "New York's Finest."

93. *Daredevil*, "New York's Finest."

94. *Daredevil*, "New York's Finest." The superhero "status" of Castle remains unclear, with fans and commentators weighing in on both sides. And in the end, it is perhaps not necessary to arrive at a consensus, for the debate itself gives the audience pause to question their own ethical choices in the treatment of others.

95. Jeff McMahan, *The Ethics of Killing* (Oxford: Oxford University Press, 2002).

96. McMahan, *The Ethics of Killing*, 242.

97. James Brabazon, *Albert Schweitzer: A Biography* (Syracuse, NY: Syracuse University Press, 2000); Mike W. Martin, *Albert Schweitzer's Reverence for Life: Ethical Idealism and Self-Realization* (Aldershot, UK: Ashgate, 2012); Nils Ole Oermann, *Albert Schweitzer: A Biography* (Oxford: Oxford University Press, 2016); and James Carleton Paget and Michael J. Thate, eds., *Albert Schweitzer in Thought and Action: A Life in Parts* (Syracuse, NY: Syracuse University Press, 2016).

98. Schweitzer, *The Philosophy of Civilization*, 309.

99. Albert Schweitzer, *The Philosophy of Civilization*, trans. C. T. Campion (Amherst, NY: Prometheus Books, 1987), 309.

100. Marvin Meyer, "Affirming Reverence for Life," in *Reverence for Life: The Ethics of Albert Schweitzer for the Twenty-First Century*, ed. Marvin Meyer and Kurt Bergel (Syracuse, NY: Syracuse University Press, 2002), 23; and John R. Everett, "Albert Schweitzer and Philosophy," *Social Research* 33, no. 4 (1966): 522.

101. Barsam, *Reverence for Life*, x.

102. Schweitzer, *The Philosophy of Civilization*, 282.

103. Meyer, "Affirming Reverence for Life," 23.

104. Barsam, *Reverence for Life*, 27.

105. Schweitzer, *The Philosophy of Civilization*, 312.

106. Schweitzer, *The Philosophy of Civilization*, 316–17.

107. Schweitzer, *The Philosophy of Civilization*, 317.

108. Barsam, *Reverence for Life*, 29.

109. *Hawkeye*, Season 1, Episode 14, "Partners, Am I Right?"

110. Schweitzer, *The Philosophy of Civilization*, 317.

111. Schweitzer, *The Philosophy of Civilization*, 319.

112. Interestingly, Vision captures the ethical concerns expressed by Dixon on the eradication of the small pox virus.

113. Schweitzer, *The Philosophy of Civilization*, 79.

114. Coogan, *Superhero*, 31.

115. Meyer, "Affirming Reverence for Life," 36.

116. Barsam, *Reverence for Life*, 38.

117. Judith Butler, *Precarious Life: The Powers of Mourning and Violence* (New York: Verso, 2004), 20.

118. Butler, *Precarious Life*, 21; Timothy Beal, *When Time is Short: Finding our Way in the Anthropocene* (Boston, MA: Beacon Press, 2022), 104.

119. Beal, *When Time is Short*, 11.

120. Joshua Trey Barnett, *Mourning in the Anthropocene: Ecological Grief and Earthly Coexistence* (East Lansing: Michigan State University Press, 2022), 13.

121. Niklas Döbler and Marius Raab, "Thinking ET: A Discussion of Exopsychology," *Acta Astronautica* 189 (2021): 699.

CHAPTER TWO: MARVEL'S MONSTROUS OTHER

1. P. Andrew Miller, "Mutants, Metaphor, and Marginalism: What X-actly Do the X-Men Stand For?" *Journal of the Fantastic in the Arts* 13, no. 3 (2003): 282–90; John M. Trushell, "American Dreams of Mutants: The X-Men—'Pulp Fiction, Science Fiction, and Superheroes," *The Journal of Popular Culture* 38, no. 1 (2004): 149–68; Neil Shyminsky, "Mutant Readers, Reading Mutants: Appropriation, Assimilation, and the X-Men," *International Journal of Comic Arts* 8, no. 2 (2006): 387–405; Jeffrey J. Kripal, *Mutants and Mystics: Science Fiction, Superhero Comics, and the Paranormal* (Chicago: University of Chicago Press, 2011); Joseph J. Darowski, *X-Men and the Mutant Metaphor: Race and Gender in the Comic Books* (Lanham, MD: Rowman & Littlefield, 2014); Joseph J. Darowski, "When Business Improved Art: The 1975 Relaunch of Marvel's Mutant Heroes," in *The Ages of the X-Men: Essays on the Children of the Atom in Changing Times*, ed. Joseph J. Darowski (Jefferson, NC: McFarland & Co., 2014); Eric Garneau and Maura Foley, "Grant Morrison's Mutants and the Post-9/11 Culture of Fear," in *The Ages of the X-Men: Essays on the Children of the Atom in Changing Times*, ed. Joseph J. Darowski (Jefferson, NC: McFarland and Co., 2014); Claudia Bucciferro, "Mutancy, Otherness, and Empathy in the X-Men," in *The X-Men Films: A Cultural Analysis*, ed. Claudia Bucciferro (Lanham, MD: Rowman & Littlefield, 2016); Ramzi Fawaz, *The New Mutants: Superheroes and the Radical Imagination of American Comics* (New York: New York University Press, 2016); Russell Blackford, *Science Fiction and the Moral Imagination: Visions, Minds, Ethics* (Cham, Switzerland: Springer, 2017); Gregory S. Parks and Matthew W. Hughey, "'A Choice of Weapons': The X-Men and the Metaphor for Approaches to Racial Equality," *Indiana Law Journal* 92, no. 5 (2017): 1–26; and Beatriz Domínguez-García, "Of Mutants and Monsters: A Posthuman Study of Verhoeven's and Wiseman's *Total Recall*," *Revista Hélice*, 7, no. 1 (2021): 37–51.

2. Olaf Stapledon, *Odd John: A Story Between Jest and Earnest* (London: Methuen, 1935).

3. Russell Blackford, *Science Fiction and the Moral Imagination* (Cham, Switzerland: Springer, 2017), 150–51.

NOTES

4. Gregory Claeys, *Dystopia: A Natural History* (Oxford: Oxford University Press, 2017), 17.

5. Barbara Creed, *Phallic Panic: Film, Horror and the Primal Uncanny* (Carlton, Australia: Melbourne University Press, 2005), viii.

6. Judith Halberstam, *Skin Shows: Gothic Horror and the Technology of Monsters* (Durham, NC: Duke University Press, 1995), 2.

7. Stephen T. Asma, *On Monsters: An Unnatural History of Our Worst Fears* (New York: Oxford University Press, 2009), 13.

8. Claeys, *Dystopia*, 58.

9. Marina Levina and Diem-My T. Bui, "Introduction: Toward a Comprehensive Monster Theory in the 21st Century," in *Monster Culture in the 21st Century: A Reader*, ed. Marina Levina and Diem-My T. Bui (London: Bloomsbury, 2013), 1.

10. Donna McCormack, "Monster Talk: A Virtual Roundtable with Mark Bould, Liv Bugge, Surekha Davies, Margrit Shildrick, and Jeffery Weinstock," *Somatechnics* 8, no. 2 (2018): 250.

11. John Douard and Pamela D. Schultz, *Monstrous Crimes and the Failure of Forensic Psychiatry* (New York: Springer, 2013), 1.

12. Claeys, *Dystopia*, 67.

13. Julia Kristeva, *Powers of Horror: An Essay on Abjection*, trans. Leon S. Roudiez (New York: Columbia University Press, 1982), 4.

14. Kristeva, *Powers of Horror*, 8.

15. Amartya Sen, *Identity and Violence: The Illusion of Destiny* (New York: W. W. Norton & Company, 2006), 1–2.

16. Stuart C. Aitken, *Family Fantasies and Community Space* (New Brunswick, NJ: Rutgers University Press, 1998), 133–34.

17. David Sibley, *Geographies of Exclusion: Society and Difference in the West* (London: Routledge, 1995), 3.

18. Sen, *Identity and Violence*, 2.

19. Sen, *Identity and Violence*, 2.

20. Kimberlé Crenshaw, "Mapping the Margins: Intersectionality, Identity Politics, and Violence against Women of Color," *Stanford Law Review* 43, no. 6 (1991): 1241–99; Sumi Cho, Kimberlé Crenshaw, and Leslie McCall, "Toward a Field of Intersectionality Studies: Theory, Applications, and Praxis," *Signs* 38, no. 4 (2013): 785–810; and Elena Ruíz, "Framing Intersectionality," in *The Routledge Companion to Philosophy of Race*, ed. Paul C. Taylor, Linda Martín Alcoff, and Luvell Anderson (New York: Routledge, 2017).

21. Michel Foucault, *The Archaeology of Knowledge and the Discourse on Language*, trans. A. M. Sheridan Smith (New York: Pantheon Books, 1972), 17.

22. Dean Spade, *Normal Life: Administrative Violence, Critical Trans Politics, and the Limits of Law* (Brooklyn, NY: South End Press, 2011), 29.

23. Spade, *Normal Life*, 30.

24. Spade, *Normal Life*, 32.

25. Quoted in Douglas Martin, "The X-Men Vanquish America," *The New York Times*, August 21, 1994: 21, 27.

26. Marvel's incorporation of mutants was not the first attempt to challenge bigotry and prejudice. In 1966, for example, Marvel introduced T'Challa, the Black Panther, the first

mainstream African American superhero; three years later, he was joined by Sam Wilson (the Falcon) as a partner for Captain America. In 1973, Marvel debuted the Chinese Shang-Chi, followed in 1975 with the Latino Hector Ayala (White Tiger) and then, in 1981, with the first Latina superheroine, Bonita Juarez (Firebird). In 1992, Northstar was revealed as the first openly gay superhero. Unfortunately, these characters were usually minor figures and/or presented as demeaning caricatures. And, in a similar way, Marvel's depiction of mutants has not always succeeded. Carolyn Cocca, for example, acknowledges that the "mutant metaphor" has served as a vehicle to critique and condemn all types of discrimination and alienation. That said, despite widespread perceptions about the racial, ethic, and gender diversity of the X-Men, Marvel's mutants have never actually been as diverse as assumed. Indeed, the X-Men have routinely been dominated by able-bodied, cis-gendered, heterosexual white men. Thus, argues Neil Shyminsky, while Marvel's mutants seemingly promote "a progressive politics of inclusion and tolerance," appearances can be deceiving. Given that many of the more popular mutants in Marvel Comics are able-bodied, cis-gendered, heterosexual white men, there is a danger that the mutant metaphor might actually reinforce inequality. For Shyminsky, Marvel's mutants, and the X-Men in particular, "allows these sorts of white men to claim oppression and a victim status even as they continue to enjoy the privilege of white male power." The marginalization of "minority" superheroes continues, albeit to a lesser extent, in more recent phases of the MCU. See Carolyn Cocca, "Containing the X-Women: De-Powering and De-Queering Female Characters," in *The X-Men Films: A Cultural Analysis*, ed. Claudia Bucciferro (Lanham, MD: Rowman & Littlefield, 2016), 80; Jeffrey A. Brown, *Panthers, Hulks, and Ironhearts: Marvel, Diversity, and the 21st Century Superhero* (New Brunswick, NJ: Rutgers University Press, 2021), 5–6; Darowski, *X-Men*, 2; and Shyminsky, "Mutant Readers," 388, 392.

27. Kim Newman, "Mutants and Monsters," in *It Came from the 1950s!: Popular Culture, Popular Anxieties*, ed. Darryl Jones, Elizabeth McCarthy, and Bernice M. Murphy (New York: Palgrave Macmillan, 2011), 55.

28. Thomas Bacon, "It's Time To Accept Agents of S.H.I.E.L.D. Isn't MCU Canon," *Screenrant*, June 13, 2021, screenrant.com/agents-shield-marvel-tv-not-mcu-canon/; Nicholas Brooks, "MCU Theory: Secret Invasion Explains Why Agents Of S.H.I.E.L.D. Isn't Canon," *CBR.com*, September 14, 2022, cbr.com/secret-invasion-agents-of-shield-canon-disney-plus/; Matt Morrison, "Is Agents Of S.H.I.E.L.D. MCU Canon? Disney CEO Revives 10-Year-Old Marvel Debate," *Screenrant*, July 16, 2023, screenrant.com/agents-of-shield-mcu-canon-bob -marvel-tv/; and Lewis Glazebrook, "Agents of S.H.I.E.L.D.'s MCU Canon Status Will Finally Be Clarified 3 Years After The Show Ended," *Screenrant*, September 7, 2023, screenrant.com/ agents-of-shield-canon-status-clarified-by-marvel/ (all accessed September 8, 2023).

29. Amy Chambers, "You Gotta Make Way for the Homo Superior: Mutation, Evolution, and Super Powers on Screen," May 30, 2018, amycchambers.com/2018/05/30/mutation -evolution-and-super-powers-on-screen/ (accessed May 11, 2023).

30. Throughout the MCU, the Avengers understand that they are frequently viewed by "ordinary" humans as being monstrous and, in turn, many of the Avengers come to see themselves as monstrous. Bruce Banner (The Hulk) is notable, but so also do nonenhanced Avengers, such as Natasha Romanoff (the Black Widow), self-identify as being monsters. Markedly, Romanoff's "confession" of being a monster in *Avengers: Age of Ultron* (2015) has received widespread condemnation and remains subject to intense debate. During an

intimate, introspective scene with Banner, Romanoff reflects on her violent past. In part, Romanoff attempts to reconcile the accusations leveled against her in *The Avengers* (2012), when Loki accused the Black Widow of having her ledger dripping with blood. "You lie and kill in the service of liars and killers," Loki charged, "You pretend to be separate, to have your own code. Something that makes up for the horrors. But they are a part of you, and they will never go away." Three years later, Banner and Romanoff are becoming emotionally close, but Banner is hesitant. As Marvel's fictional Dr. Jekyll and Mr. Hyde, Banner struggles with his other self, the monster that lies within. Romanoff then explains that during her training in the Red Room to be a Black Widow—that is, a trained assassin—she was sterilized. Romanoff explains, "You know what my final test was in the Red Room? They sterilize you. It's efficient. One less thing to worry about, the one thing that might matter more than a mission. It makes everything easier—even killing. You still think you're the only monster on the team?" For many fans and critics, Romanoff's self-identification as being monstrous was deeply disturbing, for her admission implied that her inability to reproduce reduced her to that of a monster. As Meredith Woerner and Katharine Trandocasta write, "It's not the loss of innocence through killing or being forced to live a life of betraying people. The greatest loss is motherhood." They conclude, "instead of wading into the 'red ledger' of a complicated person who did seriously heinous acts and is trying desperately to buy redemption with good deeds, we get the character who feels ruined by her barren womb." Subsequent commentators have challenged this interpretation, although reactions remain divided. See Meredith Woerner and Katharine Trendacosta, "Black Widow: This Is Why We Can't Have Nice Things," *Gizmodo*, May 5, 2015, gizmodo.com/black-widow-this-is-why-we-can-t-have-nice-things-1702333037; Alyssa Rosenberg, "The Strong Feminism Behind Black Widow, and Why the Critiques Don't Stand Up," *The Washington Post*, May 5, 2015, washingtonpost.com/news/act-four/wp/2015/05/05/black-widows-feminist-heroism/; Emily St. James, "A Guide to the Growing Controversy over Joss Whedon's Avengers and Marvel's Gender Problem," *Vox*, May 11, 2015, www.vox.com/2015/5/11/8582081/avengers-age-of-ultron-joss-whedon; Sam Adams, "How Black Widow Corrects for the Marvel Movies' Most Controversial Scene," *Slate*, July 10, 2021, slate.com/culture/2021/07/black-widow-monster-joss-whedon-ultron-controversy.html; Brian Cronin, "Was the MCU's Most Controversial Black Widow Scene Based on Marvel Comics?" *CBR.com*, January 1, 2022, cbr.com/black-widow-mcu-marvel-comics-richard-morgan-monster/ (all accessed September 9, 2023).

31. Albert Schweitzer, *The Philosophy of Civilization* (Amherst, NY: Prometheus Books, 1987), 310.

32. William H. Starbuck, "Shouldn't Organization Theory Emerge from Adolescence?" *Organization* 10, no. 3 (2003): 439.

33. Starbuck, "Shouldn't Organization Theory," 439.

34. Matthew S. Hull, "Documents and Bureaucracy," *Annual Review of Anthropology* 41 (2012): 253.

35. Craig Robertson, "'You Lie!' Identity, Paper, and the Materiality of Information," *The Communication Review* 17 (2014): 69.

36. Hull, "Documents and Bureaucracy," 257.

NOTES

37. Katharine Meehan, Ian G. R. Shaw, and Sallie A. Marston, "Political Geographies of the Object," *Political Geography* 33 (2013): 2.

38. Hull, "Documents and Bureaucracy," 259.

39. Marie-Andrée Jacob, "Form-Made Persons: Consent Forms as Consent's Blind Spot," *PoLAR: Political and Legal Anthropology Review* 30, no. 2 (2007): 251.

40. Hull, "Documents and Bureaucracy," 259.

41. Ian Hacking, "Making Up People," in *Reconstructing Individualism: Autonomy, Individuality, and the Self in Western Thought*, ed. Thomas C. Heller, Morton Sosna, and David E. Wellbery (Palo Alto, CA: Stanford University Press, 1986); Christine Holmberg, Christine Bischof, and Susanne Bauer, "Making Predictions: Computing Populations," *Science, Technology, & Human Values* 38, no. 3 (2013): 398–420; Alice Street, "Seen by the State: Bureaucracy, Visibility and Governmentality in a Papua New Guinean Hospital," *The Australian Journal of Anthropology* 23 (2012): 1–21; and Louise Amoore, *Cloud Ethics: Algorithms and the Attributes of Ourselves and Others* (Durham, NC: Duke University Press, 2020).

42. Spade, *Normal Life*, 38.

43. Spade, *Normal Life*, 38.

44. Michel Foucault, "About the Concept of the 'Dangerous Individual' in 19th-Century Legal Psychiatry," *International Journal of Law and Psychiatry* 1, no. 1 (1978): 1–18; John Pratt, "Governing the Dangerous: An Historical Overview of Dangerous Offender Legislation," *Social & Legal Studies* 5, no. 1 (1996): 21–26; Bill Hebenton and Toby Seddon, *British Journal of Criminology* 49, no. 3 (2009): 343–62; Johannes Scheu, "Dangerous Classes: Tracing Back and Epistemological Fear," *Distinktion: Scandinavia Journal of Social Theory* 12, no. 2 (2011): 115–34; and Angel Aedo and Paulina Faba, "Rethinking Prevention as a Reactive Force to Contain Dangerous Classes," *Anthropological Theory* 22, no. 3 (2022): 338–61.

45. Brian Steels, "Forever Guilty: Convict Perceptions of Pre and Post Conviction," *Current Issues in Criminal Justice* 21, no. 2 (2009): 242–56.

46. Marieke de Goede, "Fighting the Network: A Critique of the Network as a Security Technology," *Distinktion: Scandinavian Journal of Social Theory* 13, no. 2 (2012): 217.

47. Marieke de Goede and Samuel Randalls, "Precaution, Preemption: Arts and Technologies of the Actionable Future," *Environment and Planning D: Society and Space* 27, no. 5 (2009): 861.

48. Adrian Acu, "Time to Work for a Living: The Marvel Cinematic Universe and the Organized Superhero," *Journal of Popular Film and Television* 44, no. 4 (2016): 195–205; Jennifer Beckett, "Acting with Limited Oversight: S.H.I.E.L.D. and the Role of Intelligence and Intervention in the Marvel Cinematic Universe," in *Assembling the Marvel Cinematic Universe: Essays on the Social, Cultural and Geopolitical Domains*, ed. Julian C. Chambliss, William L. Svitavsky, and Daniel Fandino (Jefferson, NC: McFarland and Company, 2018); Samira Shirish Nadkarni, "'To be the Shield': American Imperialism and Explosive Identity Politics in *Agents of S.H.I.E.L.D.*," in *Assembling the Marvel Cinematic Universe: Essays on the Social, Cultural and Geopolitical Domains*, ed. Julian C. Chambliss, William L. Svitavsky, and Daniel Fandino (Jefferson, NC: McFarland and Company, 2018); Brett Pardy, "The Militarization of Marvel's Avengers," *Studies in Popular Culture* 42, no. 1

(2019): 103–22; and Juan Medina-Contreras and Pedro Sangro Colón, "Representations of Defense Organizations in the Marvel Cinematic Universe (2008–2019)," *Communication & Society* 33, no. 4 (2020): 19–32.

49. Darren Franich, "SHIELD: 10 Important Facts about Marvel's Superspy Organization," *Entertainment Weekly*, September 24, 2013, ew.com/article/2013/09/24/shield-agents-marvel/ (accessed May 21, 2023).

50. *Agents of S.H.I.E.L.D.*, Season 1, Episode 1, "Pilot."

51. Marvel's imagined intelligence agency is modeled, on the one hand, after an assemblage of (mostly) US intelligence agencies, including the CIA, the FBI, and the NSA and, on the other hand, a collage of 1960s fictional counterparts, including organizations like U.N.C.L.E. in *The Man from U.N.C.L.E.* and its antagonist T.H.R.U.S.H. That said, while S.H.I.E.L.D. is portrayed as an organization which works for the United Nations—the cinematic version answers to a shadowy 'World Security Council'—it is decidedly an American operation. See Darren Franich, "SHIELD: 10 Important Facts about Marvel's Superspy Organization," *Entertainment Weekly*, September 24, 2013, ew.com/article/2013/09/24/shield-agents-marvel/ (accessed May 21, 2023).

52. *Agents of S.H.I.E.L.D.*, Season 1, Episode 22, "Beginning of the End."

53. Spade, *Normal Life*, 24.

54. Spade, *Normal Life*, 24.

55. Louise Amoore, "Lines of Sight: On the Visualization of Unknown Futures," *Citizenship Studies* 13, no. 1 (2009): 17–30; Marieke de Goede and Gavin Sullivan, "The Politics of Security Lists," *Environment and Planning D: Society and Space* 34, no. 1 (2016): 67–88; Anna Leander, "The Politics of Whitelisting: Regulatory Work and Topologies in Commercial Security," *Environment and Planning D: Society and Space* 34, no. 1 (2016): 48–66; and Urs Stäheli, "Indexing—the Politics of Invisibility," *Environment and Planning D: Society and Space* 34, no. 1 (2016): 14–29.

56. Cornelia Vismann, *Files: Law and Media Technology*, trans. Geoffrey Winthrop-Young (Palo Alto, CA: Stanford University Press, 2008), 6.

57. Stäheli, "Indexing," 14.

58. Stäheli, "Indexing," 14.

59. de Goede and Sullivan, "Politics of Security Lists," 72.

60. de Goede and Sullivan, "Politics of Security Lists," 70.

61. de Goede and Sullivan, "Politics of Security Lists," 69.

62. de Goede and Sullivan, "Politics of Security Lists," 69. See also Daniel J. Steinbock, "Designating the Dangerous: From Blacklists to Watch Lists," *Seattle University Law Review* 30 (2006): 65–118; and Anya Bernstein, "The Hidden Costs of Terrorist Watch Lists," *Buffalo Law Review* 61, no. 3 (2013): 461–535.

63. de Goede and Sullivan, "Politics of Security Lists," 72.

64. *Captain America: Civil War* (2012).

65. *Captain America: Civil War* (2012).

66. *Captain America: Civil War* (2012).

67. *Captain America: Civil War* (2012).

68. Darowski, "When Business Improved Art," 37.

NOTES

147

69. Quoted in Dan Bova, "In Unearthed Interview, X-Men Creator Stan Lee Reveals How Being Lazy Led to a $6 Billion Idea," *Entrepreneur*, April 13, 2022, entrepreneur.com/leadership/in-unearthed-interview-x-men-creator-stan-lee-reveals-how/424694 (accessed September 8, 2023).

70. Parks and Hughey, "'A Choice of Weapons,'" 11. Darowski, though, identifies that the earliest issues—certainly before the re-launch of the X-Men in 1975—were not consistent in promoting anti-prejudicial themes. See Darowski, "When Business Improved Art," 41.

71. Quoted in Parks and Hughey, "'A Choice of Weapons,'" 11.

72. Darowski, *X-Men and the Mutant Metaphor*, 2.

73. Darowski, "When Business Improved Art," 40.

74. Miller, Mutants, Metaphor," 283.

75. Brown, *Panthers, Hulks, and Ironhearts*, 5.

76. Fawaz, *The New Mutants*, 144.

77. David Harth, "10 Reasons Marvel's Inhumans Flopped," *CBR.com*, August 13, 2021, cbr.com/inhumans-marvel-flopped-failed/; Adam Holmes, https://www.cinemablend.com/superheroes/marvel-cinematic-universe/what-is-the-inhumans-future-in-the-mcu-lets-talk-about-it; George Chrysostomou, "Is There a Future for Marvel's Inhumans in the MCU?" *denofgeek.com*, August 19, 2022, .denofgeek.com/tv/marvel-inhumans-mcu-future/; Jeremy Brown, "Where Agents of S.H.I.E.L.D. Fits Into Your MCU Rewatch," *CBR.com*, February 2, 2023, cbr.com/agents-of-shield-watch-order-mcu/; Lyvie Scott, "10 Years Later, Marvel is Walking Back its Most Short-Sighted Decision," *Inverse*, July 18, 2023, inverse.com/entertainment/ms-marvel-x-men-mcu (accessed September 5, 2023); and Catherine Mora, "10 Signs the MCU is Moving Away from the Inhumans Brand," *CBR.com*, September 3, 2022, www.cbr.com/mcu-inhumans-signs-moving-away/ (all accessed September 5, 2023).

78. Quoted in Justin Davis, "Marvel Studios Announces Inhumans Movie," *IGN.com*, October 28, 2014, ign.com/articles/2014/10/28/marvel-studios-announces-inhumans-movie (accessed September 5, 2023); see also Mora, "10 Signs the MCU is Moving Away."

79. Andy L. Kubai, "Kevin Feige Confirms Marvel's Inhumans Is Still Planned," *Screenrant*, May 14, 2016, screenrant.com/marvel-inhumans-movie-kevin-feige-update/; and Jason Berman, "Marvel's Kevin Feige Says Inhumans Movie Would Be 'Super Cool,'" *Screenrant*, November 4, 2016, screenrant.com/marvel-inhumans-movie-super-cool/ (all accessed September 5, 2023).

80. Marc Buxton, "Agents of S.H.I.E.L.D.: Afterlife Review," *denofgeek.com*, April 8, 2015, www.denofgeek.com/comics/agents-of-shield-afterlife-review/ (accessed September 5, 2023).

81. Sam Stone, "Agents of S.H.I.E.L.D.'s Farewell Marks the True End of Marvel Television," CBR.com, August 14, 2020, www.cbr.com/agents-of-shields-farewell-end-marvel-television/ (accessed September 5, 2023).

82. Ben Silverio, "Kevin Feige Talks Infinity Gems, Daredevil, & More Marvel Movies," *Sciencefiction.com*, November 12, 2013, sciencefiction.com/2013/11/12/kevin-feige-talks-infinity-gems-daredevil-marvel-movies/ (accessed September 8, 2023).

83. Diya Majumdar, "'There's some sort of division': Ming-Na Wen May Have Just Confirmed a Longstanding MCU vs Agents of S.H.I.E.L.D. Rumor," *MSN.com*, April 25,

NOTES

2024, msn.com/en-us/movies/news/there-s-some-sort-of-division-ming-na-wen-may
-have-just-confirmed-a-longstanding-mcu-vs-agents-of-shield-rumor/ar-AA1nECcf
(accessed May 4, 2024).

84. Kim Masters and Matthew Belloni, "Marvel Shake-Up: Film Chief Kevin Feige
Breaks Free of CEO Ike Perlmutter," *The Hollywood Reporter*, August 31, 2015, hollywoodre-
porter.com/news/general-news/marvel-shake-up-film-chief-819205/; Sam Stone, "Agents of
SHI.E.L.D's Farewell." See also John Castelle, "How Much Longer Can the Marvel Movies
Ignore Marvel TV?" *Screenrant*, November 16, 2016, screenrant.com/marvel-movies
-ignore-tv-inhumans/ (accessed September 5, 2023).

85. Sam Stone, "Marvel Studios Taking Over All Live-Action TV Projects," *CBR.
com*, September 20, 2019, cbr.com/marvel-studios-taking-over-live-action-tv/ (accessed
September 5, 2023).

86. Madeline Matsumoto-Duyan, "Why Agents of S.H.I.E.L.D. Breaking Away From
the MCU Is for the Better," *CBR.com*, August 17, 2020, cbr.com/agents-of-shield-breaking
-away-from-mcu-for-the-better/ (accessed September 5, 2023).

87. Jordan Iacobucci, "10 Marvel Shows We're Glad Got Canceled," *CBR.com*, May 15,
2023, cbr.com/marvel-shows-happy-to-see-canceled/#inhumans-2017 (accessed September
5, 2023).

88. Eric Francisco, "Kevin Feige wants you to forget Ms. Marvel is Inhuman in the
MCU," *Inverse*, May 11, 2022, inverse.com/entertainment/ms-marvel-mcu-inhuman;
Nikhil Makwana, "'They don't care about Inhumans anymore': Ms. Marvel's Mutant
Connection Has Pissed Off Comic-Book Fans, Say Kevin Feige Doesn't Know What He's
Doing Anymore," *Fandomwire*, July 13, 2022, fandomwire.com/kevin-feige-mcu-ms
-marvel-fans-inhumans/; and Rohit Rajput, "'Kevin Feige Hates Inhumans': Memes
Galore Following Kamala Khan's Mutant Origin Revelation in MCU's Ms. Marvel Finale,"
Sportskeeda, July 16, 2022, sportskeeda.com/comics/kevin-feige-hates-inhumans-memes
-galore-following-kamala-khan-s-mutant-origin-revelation-mcu-s-ms-marvel-finale
(accessed September 5, 2023).

89. Jamie Lovett, "Marvel Announces 'Death of the Inhumans,'" *Comicbook.com*,
March 28, 2018, comicbook.com/marvel/news/death-of-the-inhumans/ (accessed
September 5, 2023).

90. Mora, "10 Signs the MCU is Moving Away."

91. McSweeney, *Avengers Assemble*, 212.

92. Beckett, "Acting with Limited Oversight," 210.

93. Thomas Bacon, "Agents of S.H.I.E.L.D. Proved Inhumans Don't Work - By Turning
Them into Mutants," *Screenrant*, April 11, 2022, screenrant.com/agents-shield-inhumans
-xmen-mutants-mcu-marvel-bad/ (accessed September 8, 2023).

94. *Agents of S.H.I.E.L.D.*, Season 2, Episode 12, "Who You Really Are."

95. *Agents of S.H.I.E.L.D.*, Season 2, Episode 10, "What They Become."

96. *Agents of S.H.I.E.L.D.*, Season 3, Episode 1, "Laws of Nature."

97. *Agents of S.H.I.E.L.D.*, Season 2, Episode 16, "Afterlife."

98. *Agents of S.H.I.E.L.D.*, Season 3, Episode 2, "Purpose in the Machine."

99. *Agents of S.H.I.E.L.D.*, Season 3, Episode 3, "A Wanted (Inhu)Man."

100. *Agents of S.H.I.E.L.D.*, Season 3, Episode 12, "The Inside Man."

NOTES

101. *Agents of S.H.I.E.L.D.*, "Afterlife."

102. For an exceptional critique of the Orientalist tropes of Afterlife, see Nadkarni, "'To Be the Shield,'" 225–26.

103. *Agents of S.H.I.E.L.D.*, "Afterlife."

104. *Agents of S.H.I.E.L.D.*, Season 2, Episode 8, "The Things We Bury."

105. Jiaying's powers come with a price, for she must absorb the life from others to be able to regenerate. However, every few decades an elderly resident volunteers to sacrifice their life—an act of euthanasia—so Jiaying can continue living and helping others.

106. In Season 7, it is revealed that Daisy has an older sister, Kora.

107. *Agents of S.H.I.E.L.D.*, "What They Become."

108. *Agents of S.H.I.E.L.D.*, Season 4, Episode 22, "World's End."

109. *Agents of S.H.I.E.L.D.*, Season 1, Episode 5, "Girl in the Flower Dress."

110. *Agents of S.H.I.E.L.D.*, Season 1, Episode 9, "Repairs"; *Agents of S.H.I.E.L.D.*, Season 2, Episode 13, "One of Us."

111. *Agents of S.H.I.E.L.D.*, Season 2, Episode 1, "Shadows."

112. *Agents of S.H.I.E.L.D.*, Season 3, Episode 7, "Chaos Theory."

113. *Agents of S.H.I.E.L.D.*, Season 3, Episode 20, "Emancipation."

114. *Agents of S.H.I.E.L.D.*, "Pilot."

115. *Agents of S.H.I.E.L.D.*, Season 2, Episode 11, "Aftershocks."

116. *Agents of S.H.I.E.L.D.*, "Who You Really Are."

117. *Agents of S.H.I.E.L.D.*, Season 4, Episode 9, "Broken Promises."

118. *Agents of S.H.I.E.L.D.*, Season 4, Episode 10, "The Patriot."

119. *Agents of S.H.I.E.L.D.*, Season 2, Episode 20, "Scars."

120. Nadkarni, "'To Be the Shield,'" 225–26.

121. *Agents of S.H.I.E.L.D.*, "Scars."

122. *Agents of S.H.I.E.L.D.*, Season 2, Episode 21, "S.O.S. Part 1."

123. Nadkarni, "'To Be the Shield,'" 226.

124. *Agents of S.H.I.E.L.D.*, "S.O.S. Part 1."

125. *Agents of S.H.I.E.L.D.*, Season 2, Episode 22, "S.O.S. Part 2."

126. *Agents of S.H.I.E.L.D.*, "S.O.S. Part 2."

127. Foucault, *Abnormal*, 47.

128. Foucault, *Abnormal*, 44.

129. Foucault, *Abnormal*, 46.

130. Foucault, *Abnormal*, 25.

131. Foucault, *Abnormal*, 50.

132. *Agents of S.H.I.E.L.D.*, "A Wanted (Inhu)Man."

133. *Agents of S.H.I.E.L.D.*, Season 2, Episode 6, "A Fractured House."

134. *Agents of S.H.I.E.L.D.*, Season 4, Episode 3, "Uprising."

135. *Agents of S.H.I.E.L.D.*, "Uprising."

136. *Agents of S.H.I.E.L.D.*, Season 3, Episode 1, "Laws of Nature."

137. *Iron Man 3* (2013).

138. *Agents of S.H.I.E.L.D.*, "Laws of Nature."

139. *Agents of S.H.I.E.L.D.*, "Laws of Nature."

140. *Agents of S.H.I.E.L.D.*, "Laws of Nature."

141. *Agents of S.H.I.E.L.D.*, "Laws of Nature."

142. *Agents of S.H.I.E.L.D.*, "Laws of Nature."

143. *Agents of S.H.I.E.L.D.*, "Laws of Nature."

144. *Agents of S.H.I.E.L.D.*, "Laws of Nature."

145. *Agents of S.H.I.E.L.D.*, Season 3, Episode 6, "Among Us Hide. . . ."

146. Garner is also the ex-husband of S.H.I.E.L.D. agent Melinda May and later undergoes Terrigenesis, transforming into the Inhuman known as Lash.

147. *Agents of S.H.I.E.L.D.*, "S.O.S. Part 2."

148. *Agents of S.H.I.E.L.D.*, Season 3, Episode 2, "Purpose in the Machine."

149. *Agents of S.H.I.E.L.D.*, "Emancipation."

150. *Agents of S.H.I.E.L.D.*, "Emancipation."

151. Philips, "Violence in the American Imaginary," 472.

152. Philips, "Violence in the American Imaginary," 472.

153. Henderson, "Daredevil," 170.

154. Henderson, "Daredevil," 148.

155. *Agents of S.H.I.E.L.D.*, "Broken Promises."

156. *Agents of S.H.I.E.L.D.*, Season 3, Episode 12, "The Inside Man."

157. *Agents of S.H.I.E.L.D.*, Season 3, Episode 14, "Watchdogs."

158. *Agents of S.H.I.E.L.D.*, Season 4, Episode 13, "Boom."

159. *Agents of S.H.I.E.L.D.*, Season 4, Episode 12, "Hot Potato Soup."

160. *Agents of S.H.I.E.L.D.*, Season 4, Episode 14, "The Man Behind the Shield."

161. *Agents of S.H.I.E.L.D.*, "The Man Behind the Shield."

162. *Agents of S.H.I.E.L.D.*, "Watchdogs."

163. *Agents of S.H.I.E.L.D.*, "Watchdogs."

164. *Agents of S.H.I.E.L.D.*, "Watchdogs."

165. *Agents of S.H.I.E.L.D.*, "Boom,"; *Agents of S.H.I.E.L.D.*, "Broken Promises."

166. *Agents of S.H.I.E.L.D.*, Season 4, Episode 5, "Lockup."

167. *Agents of S.H.I.E.L.D.*, "Uprising."

168. *Agents of S.H.I.E.L.D.*, "Broken Promises."

169. *Agents of S.H.I.E.L.D.*, Season 4, Episode 11, "Wake Up."

170. *Agents of S.H.I.E.L.D.*, "Broken Promises."

171. *Agents of S.H.I.E.L.D.*, "Pilot.".

172. *Agents of S.H.I.E.L.D.*, "Broken Promises."

173. Schweitzer, *The Philosophy of Civilization*, 311.

174. Claeys, *Dystopia*, 27.

175. *Agents of S.H.I.E.L.D.*, "Lockup."

176. *Agents of S.H.I.E.L.D.*, "Lockup."

177. *Agents of S.H.I.E.L.D.*, "Uprising."

178. *Agents of S.H.I.E.L.D.*, "Uprising.".

179. *Agents of S.H.I.E.L.D.*, "The Patriot."

180. McSweeney, *Avengers Assemble*, 214.

181. Darren Franich, "S.H.I.E.L.D.: 10 Important Facts about Marvel's Superspy Organization," *Entertainment Weekly*, September 24, 2013, available at ew.com/article/2013/09/24/shield-agents-marvel/ (accessed May 21, 2023).

NOTES 151

182. Ben F. Silverio, "The Story Of Damage Control (So Far) In The MCU," *Slash Film*, August 25, 2022, slashfilm.com/969447/the-story-of-damage-control-so-far-in-the-mcu/; Bradley Prom, "The MCU is Wasting its Perfect S.H.I.E.L.D. Replacement," *Screenrant*, February 25, 2023, screenrant.com/mcu-damage-control-shield-replacement-wasted/; and Kai Young, "What Is The Department Of Damage Control? MCU History & Marvel Comics Origin Explained," *Screenrant*, October 10, 2023, screenrant.com/damage-control-history-origin-explained/ (all accessed December 3, 2023).

183. *Spider-Man: Homecoming* (2017).

184. Justin Wong, "The MCU's Damage Control (So Far) in the MCU," *Slash Film*, August 25, 2022, slashfilm.com/969447/the-story-of-damage-control-so-far-in-the-muc/; Bradly Prom, "The MCU is Wasting its Perfect SHIELD Replacement," *Screenrant*, February 25, 2023, screenrant.com/mcu-damage-control-shield-replacement-wasted/; and Kai Young, "What is the Department of Damage Control? MCU History and Marvel Comics Origin Explained," *Screenrant*, October 10, 2023, available at screenrant.com/damage-control-history-origin-explained (all accessed December 3, 2023).

CHAPTER THREE: MARVEL'S MODERN PROMETHEUS

1. David Soyka, "Frankenstein and the Miltonic Creation of Evil," *Extrapolation* 33, no. 2 (1992): 167.

2. Lester D. Friedman and Allison B. Kavey, *Monstrous Progeny: A History of the Frankenstein Narratives* (New Brunswick, NJ: Rutgers University Press, 2016), 3.

3. Siobhan Lyons, *Death and the Machine: Intersections of Mortality and Robotics* (Singapore: Palgrave Macmillan, 2018); Judith A. Markowitz, *Robots that Kill: Deadly Machines and Their Precursors in Myth, Folklore, Literature, Popular Culture and Reality* (Jefferson, NC: McFarland, 2019); Jo Ann Oravec, *Good Robot, Bad Robot: Dark and Creepy Sides of Robotics, Autonomous Vehicles, and AI* (Cham, Switzerland: Springer Nature, 2022).

4. Robin R. Murphy, "The Original 'I, Robot,' Featured a Murderous Robot and the Frankenstein Complex," *Science Robotics* 7, no. 71 (2022): 1-2; see also Lyons, *Death and the Machine*, 41.

5. Michael Szollosy, "Freud, Frankenstein and our Fear of Robots: Projection in our Cultural Perception of Technology," *AI & Society* 32 (2017): 435.

6. Adam Keiper and Ari N. Schulman, "The Problem with 'Friendly' Artificial Intelligence," *The New Atlantis* (2011): 88.

7. Mark Coeckelbergh, *AI Ethics* (Cambridge, MA: The MIT Press, 2020); 60. See also David J. Gunkel, *The Machine Question: Critical Perspectives on AI, Robots, and Ethics* (Cambridge, MA: The MIT Press, 2012); David J. Gunkel, *How to Survive a Robot Invasion: Rights, Responsibility, and AI* (New York: Routledge, 2020); Mark Coeckelbergh, *The Political Philosophy of AI* (Medford, MA: Polity Press, 2022); and Mark Coeckelbergh, *Robot Ethics* (Cambridge, MA: The MIT Press, 2022).

8. Russell Blackford, *Science Fiction and the Moral Imagination: Visions, Minds, Ethics* (Cham, Switzerland: Springer, 2017), 142.

9. Gunkel, *How to Survive*, 1.

10. Keiper and Schulman, "The Problem with 'Friendly' Artificial Intelligence," 81.

11. See for example the chapters in Patrick Lin, Keith Abney, and George A. Bekey, eds., *The Ethical and Social Implications of Robotics* (Cambridge, MA: The MIT Press, 2008).

12. John M. Jordan, *Robots* (Cambridge, MA: The MIT Press, 2016), 1.

13. James Barrat, *Our Final Invention: Artificial Intelligence and the End of the Human Era* (New York: Thomas Dunne Books, 2013); James Bridle, *Ways of Being: Animals, Plants, Machines: The Search for a Planetary Intelligence* (New York: Picador, 2022).

14. Max Tegmark, *Life 3.0: Being Human in the Age of Artificial Intelligence* (New York: Vintage, 2018), 50.

15. Coeckelbergh, *AI Ethics*, 64.

16. Bridle, *Ways of Being*, 10.

17. Tegmark, *Life 3.0*, 50.

18. Bridle, *Ways of Being*, 11.

19. Barrat, *Our Final Invention*, 25. See also Herbert L. Roitblat, *Algorithms are Not Enough: Creating General Artificial Intelligence* (Cambridge, MA: The MIT Press, 2020).

20. Bruce G. Buchanan, "A (Very) Brief History of Artificial Intelligence," *AI Magazine* 26, no. 4 (2005): 53.

21. Michael Haenlein and Andreas Kaplan, "A Brief History of Artificial Intelligence: On the Past, Present, and Future of Artificial Intelligence," *California Management Review* 61, no. 4 (2019): 7.

22. John McCarthy, Marvin L. Minsky, Nathaniel Rochester, and Claude E. Shannon, "A Proposal for the Dartmouth Summer Research Project on Artificial Intelligence, August 31, 1955," *AI Magazine* 27, no. 4 (2006): 12.

23. Buchanan, "A (Very) Brief History," 54; see also Ronald R. Kline, "Cybernetics, Automata Studies, and the Dartmouth Conference on Artificial Intelligence," *IEEE Annals of the History of Computing* 33, no. 4 (2010): 5–16.

24. Coeckelbergh, *AI Ethics*, 66.

25. Haenlein and Kaplan, "A Brief History," 7.

26. Coeckelbergh, *AI Ethics*, 66.

27. Haenlein and Kaplan, "A Brief History," 7.

28. Ben Goertzel, "Artificial General Intelligence: Concept, State of the Art, and Future Prospects," *Journal of Artificial General Intelligence* 5, no. 1 (2014): 1; Gonenc Gurkaynak, Ilay Yilmaz, and Gunes Haksever, "Stifling Artificial Intelligence: Human Perils," *Computer Law & Security Review* 32 (2016): 751.

29. Tegmark, *Life 3.0*, 51; Gurkaynak, Yilmaz, and Haksever, "Stifling Artificial Intelligence," 751.

30. Gurkaynak, Yilmaz, and Haksever, "Stifling Artificial Intelligence," 751.

31. Gurkaynak, Yilmaz, and Haksever, "Stifling Artificial Intelligence," 751.

32. Andrea Roli, Johannes Jaeger, and Stuart Kauffman, "How Organisms Come to Know the World: Fundamental Limits on Artificial General Intelligence," *Frontiers in Ecology and Evolution* 9 (2022): 3.

33. Tegmark, *Life 3.0*, 52.

34. Bridle, *Ways of Being*, 51–52.

NOTES

35. Ben Goertzel, "Human-Level Artificial General Intelligence and the Possibility of a Technological Singularity: A Reaction to Ray Kurzweil's *The Singularity Is Near*, and McDermott's Critique of Kurzweil," *Artificial Intelligence* 171 (2007): 1163.

36. Pei Wang, "On Defining Artificial Intelligence," *Journal of Artificial General Intelligence* 10, no. 2 (2019): 1–37.

37. Goertzel, "Artificial General Intelligence," 2–3.

38. Ronald J. Brachman, "(AA)AI: More than the Sum of its Parts," *AI Magazine* 27, no. 4 (2006): 24.

39. Barrat, *Our Final Invention*, 73.

40. Roman V. Yampolskiy, "Analysis of Types of Self-Improving Software," in *Artificial General Intelligence: 8th International Conference, AGI 2015, Berlin, Germany, July 22–25, 2015 Proceedings*, ed. Jordi Bieger, Ben Goertzel, and Alexy Potapov (Cham, Switzerland: Springer, 2015), 385.

41. Theresia Ratih Dewi Saputri, and Seok-Won Lee, "The Application of Machine Learning in Self-Adaptive Systems: A Systematic Literature Review," *IEEE Access* 8 (2020): 205948.

42. Jens Pohl, "Artificial Superintelligence: Extinction or Nirvana," in *Proceedings of InterSymp-2015, IIAS, 27th International Conference on Systems Research, Informatics, and Cybernetics.* 2015. at digitalcommons.calpoly.edu/cgi/viewcontent .cgi?article=1083&context=arch_fac (accessed November 19, 2024).

43. Ray Kurzweil, *The Singularity is Near: When Humans Transcend Biology* (New York: Penguin, 2006); Nick Bostrom, *Superintelligence: Paths, Dangers, Strategies* (Oxford: Oxford University Press, 2014).

44. Stephen M. Omohundro, "The Nature of Self-Improving Artificial Intelligence," presented at the Singularity Summit, Palo Alto, California, September 5, 2007, citeseerx .ist.psu.edu/document?repid=rep1&type=pdf&doi=4618cbdfd7dada7f61b706e4397d4e595 2b5c9a0 (accessed September 19, 2023).

45. Barrat, *Our Final Invention*, 83.

46. Barrat, *Our Final Invention*, 96.

47. Irving John Good, "Speculations Concerning the First Ultraintelligent Machine," *Advances in Computers* 6 (1965): 33.

48. Barrat, *Our Final Invention*, 100.

49. Quoted in Robert Sparrow, "Friendly AI Will Still be Our Master. Or, Why We Should Not Want to be the Pets of Super-Intelligent Computers," *AI & Society* 39, no. 5 (2024): 2439-

50. Roman V. Yampolskiy, "On the Controllability of Artificial Intelligence: An Analysis of Limitations," *Journal of Cyber Security and Mobility* 11, no. 3 (2022): 321–404. See also Stuart Russell, *Human Compatible: Artificial Intelligence and the Problem of Control* (New York: Penguin, 2019).

51. Vernor Vinge, "Technological Singularity," Paper presented at the VISION-21 Symposium, sponsored by the NASA Lewis Research Center and the Ohio Aerospace Institute, March 30–31, 1993, available at http://cmm.cenart.gob.mx/delanda/textos/ tech_sing.pdf (accessed September 20, 2023).

52. Roman V. Yampolskiy, "On the Controllability," 379.

53. Edward Lee, "Are We Losing Control?" in *Perspectives on Digital Humanism*, ed. Hannes Werthner, Erich Prem, Edward A. Lee, and Carlo Ghezzi (Cham, Switzerland: Springer, 2022), 7.

54. Sven Nyholm, "A New Control Problem? Humanoid Robots, Artificial Intelligence, and the Value of Control," *AI and Ethics* 3, no. 4 (2023): 1229-1239. By way of example, the "responsibility gap" is a significant moral and legal problem. If machines—including robots and androids—are defined as mere technological artifacts, then it is always a human who is responsible for the (mal)function of the machine. However, if it becomes possible to assign some aspect of liability to the machine, then a certain degree of responsibility shifts to the machine itself. Some real-world examples will help clarify the possible dilemma. Consider, first, a carpenter. If she hits her thumb with a hammer, blame is understandably directed toward the carpenter. Indeed, it is seemingly absurd to attribute the carpenter's mishap to the hammer. However, let's consider a computer capable of self-learning. In this scenario, the actions of the computer are unpredictable; the human engineer has no way of knowing with any certainty how the computer will evaluate and initiate subsequent decisions. Is the human still responsible for the future actions of the computer or has some threshold been crossed, whereby the computer transforms from being a tool of human technology to a technological being in its own right? For AI ethicists, this constitutes a "responsibility gap" and is intimately associated with the broader "control problem."

55. Brian Christian, *The Alignment Problem: Machine Learning and Human Values* (New York: W. W. Norton & Company, 2020), 13.

56. Eliezer Yudkowsky, *Creating Friendly AI 1.0: The Analysis and Design of Benevolent Goal Architectures* (San Francisco: The Singularity Institute, 2001); Luke Muehlhauser and Nick Bostrom, "Why We Need Friendly AI," *Think* 36, no. 13 (2014): 41–47; Robert James M. Boyles and Jeremiah Joven Joaquin, "Why Friendly AIs Won't be *that* Friendly: A Friendly Reply to Muehlhauser and Bostrom," *AI & Society* 35, no. 2 (2020): 505–7; Barbro Fröding and Martin Peterson, "Friendly AI," *Ethics and Information Technology* 23, no. 3 (2021): 207–14; Oliver Li, "Problems with 'Friendly AI,'" *Ethics and Information Technology* 23, no. 3 (2021): 543–50; and Payman Tajalli, "AI Ethics and the Banality of Evil," *Ethics and Information Technology* 23, no. 3 (2021): 447–54

57. Yudkowsky, *Creating Friendly AI*, 2–3.

58. Rosalind W. Picard, *Affective Computing* (Cambridge, MA: The MIT Press, 1997); Aaron Sloman, "Review of *Affective Computing*," *AI Magazine* 20, no. 1 (1999): 127; Rosalind W. Picard, "Affective Computing: Challenges," *International Journal of Human-Computer Studies* 59 (2003): 55–64; Michael Muller, "Multiple Paradigms in Affective Computing," *Interacting with Computers* 16, no. 4 (2004): 759–68.

59. Shaundra B. Daily, Melva T. James, David Cherry, John J. Porter III, Shelby S. Darnell, Joseph Isaac, and Tania Roy, "Affective Computing: Historical Foundations, Current Applications, and Future Trends," in *Emotions and Affect in Human Factors and Human-Computer Interaction*, ed. Myounghoon Jeon (Cambridge, MA: Academic Press, 2017), 213.

60. Picard, *Affective Computing*, x.

61. Picard, *Affective Computing*, xi.

62. Richard Yonck, *Heart of the Machine: Our Future in a World of Artificial Emotional Intelligence* (New York: Arcade Publishing, 2020), 5.

NOTES

63. Rosalind W. Picard, "Response to Sloman's Review of *Affective Computing*," *AI Magazine* 20, no. 1 (1999): 134–37; at 137.

64. Good, "Speculations," 33.

65. Isaac Asimov, "Runaround," *Astounding Science Fiction*, March (1942): 94–103. Asimov would later add a superseding "zeroth" law: A robot may not harm humanity, or, by inaction, allow humanity to come to harm.

66. Keiper and Schulman, "The Problem with 'Friendly' Artificial Intelligence," 83.

67. See for example John Storrs Hall, *Beyond AI: Creating the Conscience of the Machine* (New York: Prometheus Books, 2007); Ugo Pagallo, *The Laws of Robots: Crimes, Contracts, and Torts* (New York: Springer, 2013).

68. Keiper and Schulman, "The Problem with 'Friendly' Artificial Intelligence," 85.

69. Gunkel, *The Machine Question*, 6.

70. Blackford, *Science Fiction*, 143.

71. There is a voluminous literature on replicants as other-than-human beings. See for example Giuliana Bruno, "Ramble City: Postmodernism and 'Blade Runner,'" *October* 41 (1987): 61–74; Joe Abbott, "The 'Monster' Reconsidered: *Blade Runner*'s Replicant as Romantic Hero," *Extrapolation* 34, no. 4 (1993): 340–50; and Debbora Battaglia, "Multiplicities: An Anthropologist's Thoughts on Replicants and Clones in Popular Film," *Critical Inquiry* 27, no. 3 (2001): 493–514.

72. For further insight into both the novel and the film, see Judith Kerman (ed.), *Retrofitting Blade Runner: Issues in Ridley Scott's Blade Runner and Philip K. Dick's Do Androids Dream of Electric Sheep?* (Madison: University of Wisconsin Press, 1997); Will Brooker (ed.), *The Blade Runner Experience: The Legacy of a Science Fiction Classic* (New York: Wallflower Press, 2005); and Timothy Shanahan, *Philosophy and Blade Runner* (New York: Palgrave Macmillan, 2014).

73. Gunkel, *The Machine Question*, 6–7.

74. Quoted in Rory Cellan-Jones, "Stephen Hawking Warns Artificial Intelligence Could End Manking," *BBC News*, December 2, 2014, christusliberat.org/wp-content/uploads/2017/10/Stephen-Hawking-warns-artificial-intelligence-could-end-mankind-BBC-News.pdf (accessed May 18, 2023).

75. Karel Čapek, *R.U.R. (Rossum's Universal Robots)*, trans. Claudia Novack (New York: Penguin Books, 2004).

76. Russell Blackford, *Science Fiction and the Moral Imagination* (Cham, Switzerland: Springer, 2017), 142.

77. Rain Liivoja, "Being More than You can Be: Enhancement of Warfighters and the Law of Armed Conflict," *Law and the Future of War Research Paper No. 1* (Queensland, Australia: University of Queensland Law School, 2020), 2.

78. Jai Galliot and Mianna Lotz (eds.), *Super Soldiers: The Ethical, Legal, and Social Implications* (Aldershot, UK: Ashgate, 2015); Heather A. Harrison Dinnis and Jann K. Kleffner, "Soldier 2.0: Military Human Enhancement and International Law," *International Law Studies* 92 (2016): 432–82; William G. Braun III, Stéfanie von Hlatky, and Kim Richard Nossal (eds.), *Developing the Super Soldier: Enhancing Military Performance* (Carlisle, PA: Strategic Studies Institute, US Army War College, 2017); Andrew Bickford, "From Idiophylaxis to Inner Armor: Imagining the Self-Armoring Soldier in the United States Military from the 1960s to Today," *Comparative Studies in Society and History* 60, no. 4

(2018): 810–38; Andrew Bickford, "'Kill-Proofing' the Soldier: Environmental Threats, Anticipation, and US Military Biomedical Armor Programs," *Current Anthropology* 60, S19 (2019): S39-S48; Andre Bickford, *Chemical Heroes: Pharmacological Supersoldiers in the US Military* (Durham, NC: Duke University Press, 2020); and Andrew Bickford, "Liquid Metal Masculinity: The New Man, Will, and US Military Pharmacological Supersoldiers," *Anthropological Quarterly* 94, no. 2 (2021): 197–223.

79. Yonck, *Heart of the Machine*, 27.

80. Robert Sparrow, "Lethal Autonomous Robots Must be Stopped in their Tracks," *The Conversation*, June 3, 2013, theconversation.com/lethal-autonomous-robots-must-be -stopped-in-their-tracks-14843 (accessed May 18, 2023).

81. Heather Roff, "Autonomous or 'Semi' Autonomous Weapons? A Distinction Without Difference," *Huffington Post* blog, January 16, 2015, huffpost.com/entry/autonomous-or-semi -autono_b_6487268 (accessed May 23, 2023); see also Bode and Huelss, "Autonomous Weapons," 397.

82. Human Rights Watch, *Losing Humanity: The Case Against Killer Robots* (Human Rights Watch 2012), 2. hrw.org/sites/default/files/reports/arms1112_ForUpload.pdf (accessed May 18, 2023).

83. Precursors to fully autonomous lethal weapons are already in use. As Rebecca Crootof documents, "Weapon systems with varying levels of autonomy and lethality have already been integrated in the armed forces of numerous states. Israel, Russia, and South Korea reportedly have autonomous weapon systems currently patrolling their borders and bases; Israel and the United Kingdom have fire-and-forget missiles which independently select and engage targets; China and Russia employ sea mines which determine when and against what to deploy torpedoes; and the United States is developing and using a host of autonomous ground, air, and sea-based weapon systems." See Human Rights Watch, *Losing Humanity*, 3; and Rebecca Crootof, "The Killer Robots are Here: Legal and Policy Implications," *Cardozo Law Review* 36 (2015): 1840.

84. Asaro, "On Banning Autonomous Weapons," 690.

85. Matthew Anzarouth, "Robots that Kill: The Case for Banning Lethal Autonomous Weapon Systems," *Harvard Political Review*, December 2, 2021, harvardpolitics.com/ robots-that-kill-the-case-for-banning-lethal-autonomous-weapon-systems/ (accessed May 18, 2023).

86. Bode and Huelss, "Autonomous Weapons," 400.

87. Juergen Altmann and Frank Sauer, "Speed Kills: Why We Need to Hit the Brakes on 'Killer Robots,'" *DuckofMinerva*, April 8, 2016, duckofminerva.com/2016/04/speed-kills -why-we-need-to-hit-the-brakes-on-killer-robots.html (accessed September 28, 2023).

88. Keiper and Schulman, "The Problem with 'Friendly' Artificial Intelligence," 81.

89. Peter W. Singer, *Wired for War: The Robotics Revolution and Conflict in the 21st Century* (New York: Penguin, 2009); Gary E. Marchant, Braden Allenby, Ronald Arkin, Edward T. Barrett, Jason Borenstein, Lyn M. Gaudet, Orde Kitrie, Patrick Lin, George R. Lucas, Richard O'Meara, and Jared Silberman, "International Governance of Autonomous Military Robots," *The Columbia Science and Technology Law Review* 12 (2011): 272–315; and Jeff McMahan, *Killing by Remote Control: The Ethics of an Unmanned Military* (Oxford: Oxford University Press, 2013.

90. Marchant et al., "International Governance," 275.

NOTES

91. Milton Leitenberg, "Deaths in Wars and Conflicts in the 20th Century," *Cornell University Peace Studies Program Occasional Paper #29* (New Haven, CT: Cornell University, 2006).

92. Benjamin Shalit, *The Psychology of Conflict and Combat* (New York: Prager, 1988); Dave Grossman, *On Killing: The Psychological Cost of Learning to Kill in War and Society* (New York: Back Bay Books, 1996); James Gilligan, *Violence: Reflections on a National Epidemic* (New York: Vintage Books, 1997); James Waller, *Becoming Evil: How Ordinary People Commit Genocide and Mass Killing* (New York: Oxford University Press, 2002); Daniel Chirot and Clark McCauley, *Why Not Kill Them All? The Logic and Prevention of Mass Political Murder* (Princeton, NJ: Princeton University Press, 2006); and Kathleen Taylor, *Cruelty: Human Evil and the Human Brain* (Oxford: Oxford University Press, 2009).

93. Chirot and McCauley, *Why Not Kill Them*, 51.

94. S. L. A. Marshall, *Men Against Fire* (Glouchester, UK: Peter Smith, 1978); G. Dyer, *War* (London: Guild Publishing, 1985).

95. Grossman, *On Killing*, 28.

96. Marc Pilisuk with Jennifer A. Rountree, *Who Benefits from Global Violence and War: Uncovering a Destructive System* (Westport, CT: Praeger Security, 2008), 34.

97. Waller, *Becoming Evil*, 245.

98. Grossman, *On Killing*, 164.

99. *Avengers: Age of Ultron* (2015).

100. *Avengers: Age of Ultron* (2015).

101. Troy Benjamin, *The Wakanda Files: A Technological Exploration of the Avengers and Beyond* (Bellevue, WA: Epic Ink, 2021), 134.

102. In a plot twist, Stern is later revealed to be a member of HYDRA.

103. *Iron Man 2* (2010).

104. *Iron Man 2* (2010).

105. *Iron Man 2* (2010).

106. Stark, in particular, is traumatized by the attack and his guilt is augmented by his confrontation with Aldrich Killian, a disgruntled scientist who, in *Iron Man 3* (2013) plans to install his own shadow government.

107. Sokovia is a fictional country in the MCU.

108. It is subsequently revealed that Wanda is a witch and her exposure to the Mind Stone awakened and augmented her powers.

109. *Avengers: Age of Ultron* (2015).

110. *Avengers: Age of Ultron* (2015). Stark's phrase is of course evocative of former British Prime Minister Neville Chamberlain's pre-World War II declaration of peace after signing the 1938 Munich Agreement with Adolph Hitler.

111. *Avengers: Age of Ultron* (2015).

112. *Avengers: Age of Ultron* (2015).

113. Goertzel, "Human-Level Artificial General Intelligence," 1164.

114. Dario Floreano, Francesco Mondada, Andres Perez-Uribe, and Daniel Roggen, "Evolution of Embodied Intelligence," in *Embodied Artificial Intelligence: International Seminar, Dagstuhl Castle, Germany, July 7–11, 2003. Revised Papers* (Berlin Heidelberg: Springer, 2004), pp. 293–311.

115. Stephane Doncieux, Nicolas Bredeche, Jean-Baptiste Mouret, and Agoston E. Eiben, "Evolutionary Robotics: What, Why, and Where to," *Frontiers in Robotics and AI* 2, no. 4 (2015): 1–18; Fernando Silva, Miguel Duarte, Luís Correia, Sancho Moura Oliveira, and Anders Lyhne Christensen, "Open Issues in Evolutionary Robotics," *Evolutionary Computation* 24, no. 2 (2016): 205–36; and Emma Hart and Léni K. Le Goff. "Artificial evolution of robot bodies and control: on the interaction between evolution, learning and culture," *Philosophical Transactions of the Royal Society B* 377, no. 1843 (2022): 1–8.

116. *Avengers: Age of Ultron* (2015).

117. *Avengers: Age of Ultron* (2015).

118. *Avengers: Age of Ultron* (2015).

119. Terence McSweeney is justifiably critical, both of the film and of Tony Stark. Of the latter, McSweeney writes, "Stark's narcissistic vision, in which he once again is the only person who can save humanity, will compel him to undertake his own ethically dubious experiments and ultimately create the film's antagonist, the malicious and advanced sentient robot known as Ultron, actions he will refuse to apologize for throughout the film." I do not dispute this interpretation. However, I want to take the critique in a different direction, for the film also underscores humanity's fear and fascination of the artificial Other. See McSweeney, *Avengers Assemble*, 190.

120. *Avengers: Age of Ultron* (2015).

121. *Avengers: Age of Ultron* (2015).

122. Čapek, *R.U.R.*, 75.

123. *Avengers: Age of Ultron* (2015). In this particularly moving scene, Vision says that there may be no way the Avengers can trust him—before proceeding to wield Thor's hammer, thus displaying his worthiness.

124. *Avengers: Age of Ultron* (2015).

125. *Avengers: Age of Ultron* (2015).

126. *Avengers: Age of Ultron* (2015).

127. *Avengers: Age of Ultron* (2015).

128. *Avengers: Age of Ultron* (2015).

129. *Agents of S.H.I.E.L.D.*, Season 4, Episode 4, "Let me Stand Next to Your Fire."

130. *Agents of S.H.I.E.L.D.*, Season 4, Episode 11, "Wake Up."

131. *Agents of S.H.I.E.L.D.*, "Wake Up."

132. *Agents of S.H.I.E.L.D.*, Season 4, Episode 9, "Broken Promises."

133. *Agents of S.H.I.E.L.D.*, Season 4, Episode 12, "Hot Potato Soup."

134. *Agents of S.H.I.E.L.D.*, "Broken Promises."

135. *Agents of S.H.I.E.L.D.*, "Broken Promises."

136. *Agents of S.H.I.E.L.D.*, "Broken Promises."

137. Calvert, "The AI and the Looking Glass," 19.

138. *Agents of S.H.I.E.L.D.*, Season 4, Episode 8, "The Laws of Inferno Dynamics."

139. *Agents of S.H.I.E.L.D.*, "Broken Promises."

140. *Agents of S.H.I.E.L.D.*, Season 4, Episode 21, "The Return."

141. *Agents of S.H.I.E.L.D.*, "Broken Promises."

142. *Agents of S.H.I.E.L.D.*, "Broken Promises."

143. *Agents of S.H.I.E.L.D.*, "Broken Promises."

144. *Agents of S.H.I.E.L.D.*, Season 5, Episode 13, "Principia."

145. *Agents of S.H.I.E.L.D.*, "Hot Potato Soup."

146. Murray Shanahan, *The Technological Singularity* (Cambridge, MA: The MIT Press, 2015), 126.

147. Robert Sparrow, "The Turing Triage Test," *Ethics and Information Technology* 6 (2004): 203–13; at 211.

148. Peter Winch, "Presidential Address: 'Eine Einstellung zur Seele,'" *Proceedings of the Aristotelian Society* 81 (1980–1981): 1–15; at 8.

149. Sparrow, "The Turing Triage Test," 211.

150. Winch, "Presidential Address," 11.

151. Masahiro Mori's 1970 article was originally published in a Japanese journal called *Energy*; an English translation appears as as Masahiro Mori, "The Uncanny Valley," *IEEE Robotics & Automation Magazine*, 19, no. 2 (2012), 98–100.

152. Mark Kingwell, *Singular Creatures: Robots, Rights, and the Politics of Posthumanism* (Montreal & Kingston: McGill-Queen's University Press, 2022), 28.

153. Sparrow, "The Turing Triage Test," 211.

154. *Agents of S.H.I.E.L.D.*, "Self Control," Season 4, Episode 15.

155. Shanahan, *The Technological Singularity*, 194.

156. *Agents of S.H.I.E.L.D.*, "The Return," Season 4, Episode 21.

157. *Agents of S.H.I.E.L.D.*, "The Return," Season 4, Episode 21.

158. *Agents of S.H.I.E.L.D.*, "Identity and Change," Season 4, Episode 17.

159. Shanahan, *The Technological Singularity*, 162.

160. *Avengers: Age of Ultron* (2015).

161. *Avengers: Age of Ultron* (2015).

162. *Avengers: Age of Ultron* (2015).

163. *Agents of S.H.I.E.L.D.*, "The Man Behind the Shield," Season 4, Episode 14.

164. *Agents of S.H.I.E.L.D.*, "The Man Behind the Shield," Season 4, Episode 14.

165. Albert Schweitzer, *The Philosophy of Civilization* (Amherst, NY: Prometheus Books, 1987), 274.

166. Roman Globokar, "The Ethics of Albert Schweitzer as an Inspiration for Global Ethics," *Studia Teologiczno-Historyczne* 40, no. 1 (2020): 55–71; at 59.

167. Schweitzer, *The Philosophy of Civilization*, 268.

CHAPTER FOUR: MARVEL'S ALIEN ENCOUNTERS

1. Russell Blackford, *Science Fiction and the Moral Imagination: Visions, Minds, Ethics* (Cham, Switzerland: Springer, 2017), 131.

2. Geoff King and Tanya Krzywinska, *Science Fiction Cinema: From Outerspace to Cyberspace* (London: Wallflower Press, 2000); Neil Badmington, *Alien Chic: Posthumanism and the Other Within* (New York: Routledge, 2004); Russell Blackford, *Science Fiction and the Moral Imagination* (Cham, Switzerland: Springer, 2017); Borbála Bökö, "Human-Alien Encounters in Science Fiction: A Postcolonial Perspective," *Acta Universitatis Sapientiae, Film and Media Studies* 16 (2019): 189–203.

NOTES

3. See for example Ulrike Küchler, Silja Maehl, and Graeme Stout (editors), *Alien Imaginations: Science Fiction and Tales of Transnationalism* (New York: Bloomsbury, 2015).

4. Badmington, *Alien Chic*, 11.

5. Christina Lord, "Facing the Science-Fictional Other: Human-Alien Contact in J. H. Rosny aîné's *Les Xipéhuz*," *Nineteenth-Century French Studies* 48, nos. 1–2 (2019–2020): 149–65; at 152.

6. Bökö, "Human-Alien Encounters," 190.

7. Gregory Claeys, *Dystopia: A Natural History* (Oxford, UK: Oxford University Press, 2017), 60.

8. *Dr. Strange* (2016).

9. Octavio Chon-Torres, "Astrobioethics: Epistemological, Astroethical, and Interplanetary Issues," in *Astrobiology: Science, Ethics, and Public Policy*, edited by Octavio A. Chon-Torres, Ted Peters, Joseph Seckback, and Richard Gordon (Malden, MA: Scrivener, 2021), 1–16; at 3.

10. Badmington, *Alien Chic*, 63; Lord, "Facing the Science-Fictional Other," 152.

11. Lord, "Facing the Science-Fictional Other," 150.

12. Patrick Lucanio, *Them or Us: Archetypal Interpretations of Fifties Alien Invasion Films* (Bloomington, IN: Indiana University Press, 1987); Elana Gomel, *Representations of the Post/Human: Monsters, Aliens and Others in Popular Culture* (New Brunswick, NJ: Rutgers University Press, 2002).

13. Jutta Weldes (editor), *To Seek Out New Worlds: Science Fiction and World Politics* (New York: Palgrave Macmillan, 2003); John Rieder, "Science Fiction, Colonialism, and the Plot of Invasion," *Extrapolation* 46, no. 3 (2005): 373–94; John Rieder, *Colonialism and the Emergence of Science Fiction* (Middletown, CT: Wesleyan University Press, 2008); Sherryl Vint, *Science Fiction* (Cambridge, MA: The MIT Press, 2021); and Michael Lechuga, *Visions of Invasion: Alien Affects, Cinema, and Citizenship in Settler Colonies* (Jackson, MS: University Press of Mississippi, 2023).

14. Rieder, "Science Fiction, Colonialism," 373.

15. Rieder, "Science Fiction, Colonialism," 375.

16. Lord, "Facing the Science-Fictional Other," 152; see also Gomel, *Science Fiction, Alien Encounters*, 45.

17. King and Krzywinska, *Science Fiction Cinema*, 31–32.

18. Blackford, *Science Fiction and the Moral Imagination*, 133.

19. *The Day the Earth Stood Still* (1951).

20. Catherine Marshall, "Refugees and Other Aliens," *Eureka Street* 19, no. 16 (2009): 38–39; Samantha Kountz, "We Come in Peace: Immigration in Post-Cold War Science Fiction Film," *Foundation* 43, no. 119 (2014): 29–40; Dina Al Awadhi and Jason Dittmer, "The Figure of the Refugee in Superhero Cinema," *Geopolitics* 27, no. 2 (2022): 604–28.

21. Marshall, "Refugees and Other Aliens," 38.

22. Paul Davies, "The Eerie Silence," *Physics World* 23, no. 3 (2010): 28–33; at 29.

23. Linda Billings, "Earth, Life, Space: The Social Construction of the Biosphere and the Expansion of the Concept into Outer Space," in *Social and Conceptual Issues in Astrobiology*, edited by Kelly C. Smith and Carlos Mariscal (Oxford, UK: Oxford University Press, 2020), 239–62; at 241.

NOTES

24. Louis N. Irwin and Dirk Schulze-Makuch, "The Astrobiology of Alien Worlds: Known and Unknown Forms of Life," *Universe* 6, no. 9 (2020): 6090130

25. Christopher F. Chyba and Kevin P. Hand, "Astrobiology: The Study of the Living Universe," *Annual Review of Astronomy and Astrophysics* 43 (2005): 33.

26. Charles S. Cockell, "Planetary Protection: A Microbial Ethics Approach," *Space Policy* 21, no. 4 (2005): 287–92; Michael Meltzer, *When Biospheres Collide: A History of NASA's Planetary Protection Programs* (Washington, DC: US Government Printing Office, 2010).

27. Joshua Lederberg, "Exobiology: Approaches to Life Beyond the Earth," *Science* 132, no. 3424 (1960): 393–400. See also Meltzer, *When Biospheres Collide*, 32. See also Billings, "Earth, Life, Space," 241.

28. Meltzer, *When Biospheres Collide*, 18. Of concern also was the threat of backward contamination, that is, the introduction of extraterrestrial life forms into the Earth's biosphere.

29. See for example Octavio A. Chon-Torres, "Disciplinary Nature of Astrobiology and Astrobioethic's Epistemic Foundations," *International Journal of Astrobiology* 20, no. 3 (2021): 186–93; and Octavio A. Chon-Torres and Konrad Szocik, "A Brief Epistemological Discussion of Astrotheology in the Light of Astrobiology," *International Journal of Astrobiology* 21, no. 1 (2022): 1–8.

30. Carl Sagan, *Cosmos* (London: Abacus, 2003), 130.

31. Alan Marshall, "Development and Imperialism in Space," *Space Policy* 11, no. 1 (1995): 41–52; Linda Billings, "To the Moon, Mars, and Beyond: Culture, Law, and Ethics in Space-Faring Societies," *Bulletin of Science, Technology, and Society* 26, no. 5 (2006): 430–37; Deondre Smiles, "The Settler Logics of (Outer) Space," *Society and Space*, October 26, 2020, www.societyandspace.org/articles/the-settler-logics-of-outer-space; and Alessandro Marino, "Astroenvironmentalism as SF: Bordering (and Ordering) Otherworldly Ecologies," *Environmental Humanities* 15, no. 1 (2023): 25–43.

32. Dirk Schulze-Makuch and Paul Davies, "Destination Mars: Colonization via Initial One-Way Missions," *Journal of the British Interplanetary Society* 66 (2013): 11–14; at 12.

33. Alessandra Marino and Thomas Cheney, "Centering Environmentalism in Space Governance: Interrogating Dominance and Authority Through a Critical Legal Geography of Outer Space," *Space Policy* 63 (2023): 6. See also Natalie B. Treviño, "The Cosmos is Not Finished," *Electronic Thesis and Dissertation Repository*. The University of Western Ontario, PhD thesis, 2020. ir.lib.uwo.ca/etd/7567.

34. Davies, "The Eerie Silence," 29.

35. Billings, "Earth, Life, Space," 240.

36. Blackford, *Science Fiction and the Moral Imagination*, 131.

37. Matt Purslow, "Secret Invasion Full Series Review," *IGN.com*, July 27, 2023, ign.com/articles/secret-invasion-full-series-review (accessed November 27, 2023).

38. Neal Curtis and Valentina Cardo, "Superheroes and Third-Wave Feminism," *Feminist Media Studies* 18, no. 3 (2018): 381–96; Carolyn Cocca, *Wonder Woman and Captain Marvel: Militarism and Feminism in Comics and Film* (New York: Routledge, 2020); Natalie Le Clue and Janelle Vermaak-Griessel, "Artificial Feminism: Fan Reaction to the Representation of Captain Marvel," in *Gender and Action Films: 2000 and Beyond*, edited by Steven Gerrard and Renee Middlemost (Leeds, UK: Emerald Publishing, 2022), pp. 73–85; Miriam Kent, "'Let's Rewrite Some History, Shall We?': Temporality and Postfeminism in *Captain*

Marvel's Comic Book Superhero(ine)ism," *Feminist Media Studies* 23, no. 2 (2023): 394–410; Jessica Taylor and Laura Glitsos, "'Having it Both Ways': Containing the Champions of Feminism in Female-Led Origin and Solo Superhero Films," *Feminist Media Studies* 23, no. 2 (2023): 656–70; and Aviva Dove-Viebahn, "Fantasies of White Feminism: The Human as "Other" in *Captain Marvel*," *Feminist Media Studies* 23, no. 2 (2023): 1–15.

39. *Captain Marvel* (2019).

40. Albert Schweitzer, *The Philosophy of Civilization* (Amherst, NY: Prometheus Books, 1987), 309.

41. *WandaVision*, Season 1, Episode 7, "Breaking the Fourth Wall."

42. *WandaVision*, Season 1, Episode 8, "Previously On."

43. *WandaVision*, Season 1, Episode 4, "We Interrupt This Program."

44. *Secret Invasion*, Season 1, Episode 2, "Promises."

45. *Secret Invasion*, Season 1, Episode 1, "Resurrection."

46. *Secret Invasion*, "Promises."

47. *Secret Invasion*, "Promises."

48. Fury does appear to be sympathetic with the Skrull; it is also revealed later that Fury secretly married a Skrull—although she is never seen in public with Fury in her own Skrull skin.

49. *Secret Invasion*, Season 1, Episode 6, "Home."

50. *Secret Invasion*, "Resurrection."

51. *Secret Invasion*, Season 1, Episode 3, "Betrayed."

52. *Secret Invasion*, "Betrayed."

53. *Secret Invasion*, "Promises."

54. *Secret Invasion*, "Promises."

55. Erik Amaya, "Secret Invasion Finale Debrief: Blowback and What Happens Next in the MCU," *Rotten Tomatoes*, July 26, 2023, editorial.rottentomatoes.com/article/secret -invasion-finale-debrief-blowback-and-what-happens-next-in-the-mcu/ (accessed December 18, 2023).

56. *Secret Invasion*, "Betrayed."

57. *Secret Invasion*, "Betrayed."

58. *Secret Invasion*, "Home."

59. *Secret Invasion*, "Home."

60. *Captain Marvel* (2019).

61. *Secret Invasion*, Season 1, Episode 5, "Harvest."

62. If we acknowledge *Agents of S.H.I.E.L.D.* as canon, the Asgardian presence on Earth long predates the arrival of Thor and the impact of that encounter remains evident in cultural form. In the first season of the series, Daisy Johnson receives a history lesson from Phil Coulson. "So," Daisy Johnson asks Phil Coulson, "Asgardians are aliens from another planet that visited us thousands of years ago?" "Or more," Coulson replies. Johnson follows: "And because we couldn't understand aliens, we thought they were gods?" Nodding in agreement, Coulson explains, "That's were our Norse mythology comes from." In the same episode, Jemma Simmons provides a more poetic interpretation. In a voice-over, she reflects, "In ancient times, people believed the heavens were filled with gods and monsters and magical worlds. Then, as time passed, those beliefs faded into myth and folklore. But now, we know, the stories are true. Other worlds, with names like Asgard, do exist." See

Agents of S.H.I.E.L.D., Season 1, Episode 8, "The Well." In addition, *Agents of S.H.I.E.L.D.* establishes that the Kree visited Earth over 5,000 years ago to create the Inhumans.

63. It should not go unnoticed that a more-than-human being, Bruce Banner (the Hulk) is on board and risks his life in a desperate attempt to defend the Asgardians.

64. New Asgard was actually build on the site of Tønsberg, a settlement that many generations ago was occupied by the Asgardians.

65. Al Awadhi and Dittmer, "The Figure of the Refugee," 620.

66. Not surprisingly, Marvel's *Secret Invasion* leaves many questions unanswered. As Erik Amaya wonders, "Will any Asgardians living in the US be forced to leave as the wording of his order makes them the enemy in a conflict they had no part in? Will he attempt to take this up with the UN in order to remove New Asgard from the map? Will Peter Quill still be allowed to crash on his grandfather's couch as he is half alien?" See Erik Amaya, "Secret Invasion Finale Debrief: Blowback and What Happens Next in the MCU," *Rotten Tomatoes*, July 26, 2023, rottentomatoes.com/article/secret-invasion-finale-debrief-blowback-and-what -happens-next-in-the-mcu/ (accessed December 18, 2023).

67. Gary Westfahl, ed., *Space and Beyond: The Frontier Theme in Science Fiction* (Westport, CT: Greenwood, 2000); and Rob Kitchin and James Kneale, eds., *Lost in Space: Geographies of Science Fiction* (New York: Continuum, 2002).

68. Martyn J. Fogg, "The Ethical Dimensions of Space Settlement," *Space Policy* 16 (2000): 205–11; Lynda Williams, "Irrational Dreams of Space Colonization," *Peace Review* 22, no. 1 (2010): 4–8; James S. J. Schwartz, "On the Moral Permissibility of Terraforming," *Ethics and the Environment* 18, no. 2 (2013): 1–31; Gregory Anderson, "The Politics of Settling Space," *Futures* 110 (2019): 64–66; Linda Billings, "Should Humans Colonize Mars? No," *Theology and Science* 17, no. 3 (2019): 341–46; Linda Billings, "Colonizing Other Planets is a Bad Idea," *Futures* 110 (2019): 44–46; Lori Marino, "Humanity is Not Prepared to Colonize Mars," *Futures* 110 (2019): 15–18; Kelly Smith and Keith Abney, "Human Colonization: A World Too Far?" *Futures* 110 (2019): 1–3; Konrad Szocik, "Should and Could Humans go to Mars? Yes, But Not Now and Not in the Near Future," *Futures* 105 (2019): 54–66; John W. Traphagan, "Which Humanity Would Space Colonization Save?" *Futures* 110 (2019): 47–49; Milan M. Ćirković, "Risks of Space Colonization," *Futures* 126 (2021): 102638; Ryan Gunderson, Diana Stuart, and Brian Petersen, "In Search of Plan(et) B: Irrational Rationality, Capitalist Realism, and Space Colonization," *Futures* 134 (2021): 102857; James S. J. Schwartz, Sheri Wells-Jensen, John W. Traphagan, Deana Weibel, and Kelly Smith, "What Do We Need to Ask Before Settling Space?" *Journal of the British Interplanetary Society* 74 (2021): 2–9; Milan M. Ćirković, "The Nutshell Kings: Why is Human Space Settlement Controversial in the First Place?" *Futures* 143 (2022): 103023; Konrad Szocik, "Space Not For Everyone: The Problem of Social Exclusion in the Concept of Space Settlement," *Futures* 145 (2023): 103073; and Milan M. Ćirković, "Ballet Not For Everyone: A Reply to Szocik," *Futures* 147 (2023): 103116.

69. Schulze-Makuch and Davies, "Destination Mars," 11; Schwartz et al., "What Do We Need to Ask," 3.

70. Schwartz et al., "What Do We Need to Ask," 3.

71. Martin J. Rees, *Our Final Hour: A Scientist's Warning: How Terror, Error, and Environmental Disaster Threaten Humankind's Future in this Century—On Earth and Beyond* (New York: Basic Books, 2003); at 8.

164 NOTES

72. Nick Bostrom, "Existential Risk Prevention as Global Priority," *Global Policy* 4, no. 1 (2013): 15–16.

73. Thomas Moynihan, *X-Risk: How Humanity Discovered its Own Extinction* (Falmouth, UK: Urbanomic, 2020); Toby Ord, *The Precipice: Existential Risk and the Future of Humanity* (New York: Hachette Books, 2020); Joe P. L. Davidson, "Extinctiopolitics: Existential Risk Studies, the Extinctiopolitical Unconscious, and the Billionaires' Exodus from Earth," *New Formations* 107/108 (2022): 48–65; Noah B. Taylor, *Existential Risks in Peace and Conflict Studies* (Cham, Switzerland: Springer Nature, 2023); and Émile P. Torres, *Human Extinction: A History of the Science and Ethics of Annihilation* (New York: Routledge, 2024).

74. Matt Boyd and Nick Wilson, "Existential Risk to Humanity Should Concern International Policymakers and More Could Be Done in Considering Them at the International Governance Level," *Risk Analysis* 40, no. 11 (2020): 2310.

75. Davidson, "Extinctiopolitics," 50.

76. *Avengers: Age of Ultron* (2015).

77. Davidson, "Extinctiopolitics," 48.

78. Elon Musk, "Making Humans a Multi-Planetary Species," *New Space* 5, no. 2 (2017): 46.

79. Davidson, "Extinctiopolitics," 49.

80. Traphagan, "Which Humanity," 47.

81. Hans Jonas, *The Imperative of Responsibility: In Search of an Ethics for the Technological Age* (Chicago: University of Chicago Press, 1984), 98.

82. Jonas, *The Imperative of Responsibility*, 37.

83. Gunderson et al., "In Search of Plan(et) B," 2.

84. Davidson, "Extinctiopolitics," 49.

85. Davidson, "Extinctiopolitics," 50.

86. Ben Anderson, "Preemption, Precaution, Preparedness: Anticipatory Action and Future Geographies," *Progress in Human Geography* 34, no. 6 (2010): 778.

87. Billings, "Colonizing Other Planets," 45.

88. Schwartz et al., "What Do We Need to Ask," 4.

89. See Schwartz et al., "What Do We Need to Ask," 2.

90. Williams, "Irrational Dreams," 4.

91. Ćirković, "Ballet Not for Everyone," 4.

92. Ćirković, "Ballet Not for Everyone," 3 (emphasis added).

93. Szocik, "Space Not for Everyone," 2.

94. Traphagan, "Which Humanity," 48.

CHAPTER FIVE: ENDGAMES

1. *Agents of S.H.I.E.L.D.*, "The One Who Will Save Us All," Season 5, Episode 20.

2. John M. Fischer, *Death, Immortality, and Meaning in Life* (New York: Oxford University Press, 2020), 20.

3. There is a vast literature on meaningful life; the following accounts have been helpful in shaping my understanding: Viktor E. Frankl, *Man's Search for Meaning* (Boston:

NOTES

Beacon Press, 2006); Susan Wolf, *Meaning in Life and Why it Matters* (Princeton, NJ: Princeton University Press, 2010); and Thaddeus Metz, *Meaning in Life* (Oxford, UK: Oxford University Press, 2013).

4. Fischer, *Death*, 105.

5. Fischer, *Death*, 81.

6. Martha C. Nussbaum, *The Therapy of Desire: Theory and Practice in Hellenistic Ethics* (Princeton, NJ: Princeton University Press, 2013), 229.

7. *Agents of S.H.I.E.L.D.*, "The Real Deal," Season 5, Episode 12.

8. Zygmunt Bauman, *Mortality, Immortality, and Other Life Strategies* (Palo Alto, CA: Stanford University Press, 1992), 9.

9. Olivia Stevenson, Charlotte Kenten, and Avril Maddrell, "And Now the End is Near: Enlivening and Politicizing the Geographies of Dying, Death and Mourning," *Social & Cultural Geography* 17, no. 2 (2016): 157. See also Philippe Ariès, *Western Attitudes Toward Death: From the Middle Ages to the Present.* (Baltimore: Johns Hopkins University Press, 1975); Philippe Ariès, *The Hour of Our Death* (New York, NY: Vintage, 1981); Allan Kellehear, *A Social History of Dying* (New York: Cambridge University Press, 2007); James D. E. Watson, "The Harm of Premature Death: Immortality—the Transhumanist Challenge," *Ethical Perspectives* 16, no. 4 (2009): 435–58; and Hans Ruin, *Being with the Dead: Burial, Ancestral Politics, and the Roots of Historical Consciousness* (Palo Alto, CA: Stanford University Press, 2018).

10. Bauman, *Mortality*, 4.

11. Timothy C. Baker, *Writing Animals: Language, Suffering, and Animality in Twenty-First-Century Fiction* (Cham, Switzerland: Palgrave Macmillan, 2019), 109.

12. Judith Butler, *Frames of War: When is Life Grievable?* (London, UK: Verso Books, 2009); Avril Maddrell and James D. Sidaway, eds., *Deathscapes: Spaces for Death, Dying, Mourning and Remembrance* (Aldershot, UK: Ashgate Publishing, 2012).

13. Alan Segal, *Life After Death: A History of the Afterlife in Western Religion* (New York: Doubleday, 2010).

14. Colleen McDannell and Bernhard Lang, *Heaven: A History* (New Haven, CT: Yale University Press, 2001).

15. Samuel Scheffler, *Death and the Afterlife* (New York: Oxford University Press, 2016), 68.

16. Todd May, *Death* (Durham, UK: Acumen, 2014), 36.

17. Watson, "The Harm of Premature Death," 438.

18. *Agents of S.H.I.E.L.D.*, Season 3, Episode 9, "Closure."

19. Todd May, *Death* (Durham, UK: Acumen, 2009), 4.

20. See for example Carolyn Cocca, *Superwomen: Gender, Power, and Representation* (New York: Bloomsbury Academic, 2016); Wendy Haslem, Elizabeth MacFarlane, and Sarah Richardson, eds., *Superhero Bodies: Identity, Materiality, Transformation* (New York: Routledge, 2018); Samira Shirish Nadkarni, "'To be the Shield': American Imperialism and Explosive Identity Politics in *Agents of S.H.I.E.L.D.*," in *Assembling the Marvel Cinematic University: Essays on the Social, Cultural and Geopolitical Domains*, ed. Julian C. Chambliss, William L. Svitavsky and Daniel Fandino (Jefferson, NC: MacFarland and Company, 2018); Bronwen Calvert, "The AI and the Looking Glass: Embodiment, Virtuality, and Power in

Agents of S.H.I.E.L.D., Season 4," *Slayage: The Journal of Whedon Studies* 17, no. 1 (2019): 11–35; Jeffrey A. Brown, *Panthers, Hulks and Ironhearts: Marvel, Diversity, and the 21st Century Superhero* (New Brunswicl, NJ: Rutgers University Press, 2021); and Lewis Call, "'Alien Commies from the Future!': Diversity, Equity, and Inclusion in Season Seven of *Agents of S.H.I.E.L.D.*," *Slayage: The International Journal of Whedon Studies* 19, no. 1–2 (2021): 143–84.

21. Brown, *Panthers, Hulks and Ironhearts*, 1–3.

22. Nadkarni, "'To be the Shield,'" 233–34.

23. Nadkarni, "'To Be the Shield,'" 221.

24. Terence McSweeney, *The Contemporary Superhero Film: Projections of Power and Identity* (New York: Wallflower, 2020), 9.

25. Alex S. Romagnoli and Gian S. Pagnucci, *Enter the Superheroes: American Values, Culture, and the Canon of Superhero Literature* (Lanham, MD: Scarecrow Press, 2013), 8.

26. The idea of mind-uploading has long been a staple of speculative fiction.

27. Agathe François, "The Figuration of Post-Human Bodies: A Processual Experiment with Imaginaries," *Im@go: A Journal of the Social Imaginary* 12, no. 7 (2018): 28.

28. McSweeney, *Avengers Assemble*, 137.

29. The resurrection of 'dead' characters happens for a variety of reasons within the superhero genre. From a pragmatic standpoint, real-world contract negotiations may influence whether a character returns to life or remains 'dead'—at least, temporarily. In the fictional world, however, the (apparent) death and resurrection of a 'major' character serves as an important plot device. This is seen most clearly with the 'return' of Steve Rogers/Captain America's childhood friend, Bucky Barnes/The Winter Soldier.

30. *Agents of S.H.I.E.L.D.*, Season 7, Episode 6, "Adapt or Die."

31. *Agents of S.H.I.E.L.D.*, Season 1, Episode 12, "Seeds."

32. *Iron Man 3* (2013).

33. Todd May, *A Significant Life: Human Meaning in a Silent Universe* (Chicago, IL: University of Chicago Press, 2015), 132.

34. May, *Death*, 7.

35. May, *A Significant Life*, 138.

36. May, *A Significant Life*, 94.

37. *Hawkeye*, Season 1, Episode 4, "Partners, Am I Right?"

38. May, *A Significant Life*, 136–37. (emphasis in original.)

39. *Agents of S.H.I.E.L.D.*, Season 4, Episode 15, "Self-Control." In this exchange, LMD Melinda May refers to the phenomenon experienced by persons who have limps amputated, that is, the sensation of still having the lost limb.

40. Lyons, *Death and the Machine*, 50.

41. Certainly, many nonhuman animals exhibit an awareness of possible death. However, as May explains, "to seek to remain alive when one's life is threatened. . . . is not the same as having an ongoing awareness of one's death." See May, *Death*, 7.

42. *Agents of S.H.I.E.L.D.*, Season 1, Episode 22, "Beginning of the End."

BIBLIOGRAPHY

Abram, David. *The Spell of the Sensuous: Perception and Language in a More-Than-Human World.* New York: Pantheon, 1996.

Acu, Adrian. "Time to Work for a Living: The Marvel Cinematic Universe and the Organized Superhero." *Journal of Popular Film and Television* 44, no. 4 (2016): 195–205.

Aedo, Angel, and Paulina Faba. "Rethinking Prevention as a Reactive Force to Contain Dangerous Classes." *Anthropological Theory* 22, no. 3 (2022): 338–61.

Aitken, Stuart C., *Family Fantasies and Community Space* (New Brunswick, NJ: Rutgers University Press, 1998).

AlAwadhi, Dina, and Jason Dittmer. "The Figure of the Refugee in Superhero Cinema." *Geopolitics* 27, no. 2 (2022): 604–28.

Almas, Viani Alifa. "Reconstructing the Concept of Superhero in Ms. Marvel TV Series: First Female Moslem Superhero's Journey." *KITERA KULTURA: Journal of Literary and Cultural Studies* 10, no. 3 (2022): 12–17.

Amoore, Louise. "Lines of Sight: On the Visualization of Unknown Futures." *Citizenship Studies* 13, no. 1 (2009): 17–30.

Amoore, Louise. *Cloud Ethics: Algorithms and the Attributes of Ourselves and Others.* Durham, NC: Duke University Press, 2020.

Anderson, Ben. "Preemption, Precaution, Preparedness: Anticipatory Action and Future Geographies." *Progress in Human Geography* 34, no. 6 (2010): 777–98.

Anderson, Gregory. "The Politics of Settling Space." *Futures* 110 (2019): 64–66.

Anderson, Michael, and Susan Leigh Anderson. "Machine Ethics: Creating an Ethical Intelligent Agent." *AI Magazine* 28, no. 4 (2007): 15–26.

Anderson, Michael, and Susan Leigh Anderson, eds. *Machine Ethics.* Cambridge: Cambridge University Press, 2011.

Anderson, Rani-Henrik, Boyd Cothran, and Saara Kekki, eds. *Bridging Cultural Concepts of Nature: Indigenous People and Protected Spaces of Nature.* Helsinki: Helsinki University Press, 2021.

Andrews, Gavin J., and Andrea Rishworth. "New Theoretical Terrains in Geographies of Wellbeing: Key Questions of the Posthumanist Turn." *Wellbeing, Space and Society* 4 (2023): 100–130.

Anthis, Jacy Reese, and Eze Paez. "Moral Circle Expansion: A Promising Strategy to Impact the Far Future." *Futures* 130 (2021): 102756.

Ariès, Philippe. *Western Attitudes Toward Death: From the Middle Ages to the Present.* Baltimore: Johns Hopkins University Press, 1975.

BIBLIOGRAPHY

Ariès, Philippe. *The Hour of Our Death*. New York: Vintage, 1981.

Asimov, Isaac. "Runaround." *Astounding Science Fiction* (Mar. 1942): 94–103.

Asma, Stephen T. *On Monsters: An Unnatural History of Our Worst Fears*. New York: Oxford University Press, 2009.

Badmington, Neil. *Alien Chic: Posthumanism and the Other Within*. New York: Routledge, 2004.

Baker, David. *The Shortest History of Our Universe: The Unlikely Journey from the Big Bang to Us*. New York: The Experiment, 2023.

Baker, Timothy C. *Writing Animals: Language, Suffering, and Animality in Twenty-First-Century Fiction*. Cham, Switzerland: Palgrave Macmillan, 2019.

Barnett, Joshua T. *Mourning in the Anthropocene: Ecological Grief and Earthly Coexistence*. East Lansing: Michigan State University Press, 2022.

Barrat, James. *Our Final Invention: Artificial Intelligence and the End of the Human Era*. New York: Thomas Dunne Books, 2013.

Barsam, Ara. *Reverence for Life: Albert Schweitzer's Great Contribution to Ethical Thought*. Oxford: Oxford University Press, 2008.

Bassham, Gregory. *Environmental Ethics: The Central Issues*. Indianapolis: Hackett Publishing, 2020.

Bauman, Zygmunt. *Mortality, Immortality, and Other Life Strategies*. Paolo Alto: Stanford University Press, 1992.

Beal, Timothy. *When Time is Short: Finding Our Way in the Anthropocene*. Boston: Beacon Press, 2022.

Beckett, Jennifer. "Acting with Limited Oversight: S.H.I.E.L.D. and the Role of Intelligence and Intervention in the Marvel Cinematic Universe." In *Assembling the Marvel Cinematic Universe: Essays on the Social, Cultural and Geopolitical Domains*, edited by Julian C. Chambliss, William L. Svitavsky, and Daniel Fandino. Jefferson, NC: McFarland & Co., 2018.

Benjamin, Troy. *The Wakanda Files: A Technological Exploration of the Avengers and Beyond*. Bellevue, WA: Epic Ink, 2021.

Bernat, James L. "The Biophilosophical Basis of Whole-Brain Death." *Social Philosophy and Policy* 19, no. 2 (2002): 324–42.

Bernstein, Anya. "The Hidden Costs of Terrorist Watch Lists." *Buffalo Law Review* 61, no. 3 (2013): 461–535.

Bickford, Andrew. "From Idiophylaxis to Inner Armor: Imagining the Self-Armoring Soldier in the United States Military from the 1960s to Today." *Comparative Studies in Society and History* 60, no. 4 (2018): 810–38.

Bickford, Andrew. "'Kill-Proofing' the Soldier: Environmental Threats, Anticipation, and US Military Biomedical Armor Programs." *Current Anthropology* 60, no. S19 (2019): S39-S48.

Bickford, Andrew. *Chemical Heroes: Pharmacological Supersoldiers in the US Military*. Durham, NC: Duke University Press, 2020.

Bickford, Andrew. "Liquid Metal Masculinity: The New Man, Will, and US Military Pharmacological Supersoldiers." *Anthropological Quarterly* 94, no. 2 (2021): 197–223.

BIBLIOGRAPHY

Billings, Linda. "To the Moon, Mars, and Beyond: Culture, Law, and Ethics in Space-Faring Societies." *Bulletin of Science, Technology, and Society* 26, no. 5 (2006): 430–37.

Billings, Linda. "Should Humans Colonize Mars? No." *Theology and Science* 17, no. 3 (2019): 341–46.

Billings, Linda. "Colonizing Other Planets is a Bad Idea." *Futures* 110 (2019): 44–46.

Billings, Linda. "Earth, Life, Space: The Social construction of the Biosphere and the Expansion of the Concept into Outer Space." In *Social and Conceptual Issues in Astrobiology*, edited by Kelly C. Smith and Carlos Mariscal. Oxford: Oxford University Press, 2020.

Blackford, Russell. *Science Fiction and the Moral Imagination: Visions, Minds, Ethics.* Cham, Switzerland: Springer, 2017.

Bökö, Borbála. "Human-Alien Encounters in Science Fiction: A Postcolonial Perspective." *Acta Universitatis Sapientiae, Film and Media Studies* 16 (2019): 189–203.

Bostrom, Nick. "Existential Risk Prevention as Global Priority." *Global Policy* 4, no. 1 (2013): 15–31.

Bostrom, Nick. *Superintelligence: Paths, Dangers, Strategies.* Oxford: Oxford University Press, 2014.

Boyd, Matt, and Nick Wilson. "Existential Risk to Humanity Should Concern International Policymakers and More Could Be Done in Considering Them at the International Governance Level." *Risk Analysis* 40, no. 11 (2020): 2303–12.

Boyles, Robert J. M., and Jeremiah J. Joaquin. "Why Friendly AIs Won't Be *That* Friendly: A Friendly Reply to Muehlhauser and Bostrom." *AI & Society* 35, no. 2 (2020): 505–7.

Brabrazon, James. *Albert Schweitzer: A Biography.* Syracuse, NY: Syracuse University Press, 2000.

Brachman, Ronald J. "(AA)AI): More than the Sum of Its Parts." *AI Magazine* 27, no 4 (2006): 19–34.

Braidotti, Rosi, "The Posthuman Predicament." In *The Scientific Imaginary in Visual Culture*, edited by Anneke Smelik. Goettingen, Sweden: V&R Press, 2010.

Braidotti, Rosi. *The Posthuman.* Cambridge, MA: Polity Press, 2013.

Braidotti, Rosi. *Posthuman Knowledge.* Malden, MA: Polity Press, 2019.

Braidotti, Rosi. "'We' Are in *This* Together, But We Are Not One and the Same." *Bioethical Inquiry* 17, no. 4 (2020): 465–69.

Braun III, William G., Stéfanie von Hlatky, and Kim Richard Nossal, eds. *Developing the Super Soldier: Enhancing Military Performance.* Carlisle, PA: Strategic Studies Institute, US Army War College, 2017.

Bridle, James. *Ways of Being: Animals, Plants, Machines: The Search for a Planetary Intelligence.* New York: Picador, 2022.

Brooker, Will, ed. *The Blade Runner Experience: The Legacy of a Science Fiction Classic.* New York: Wallflower Press, 2005.

Brown, Jeffrey A. *The Modern Superhero in Film and Television: Popular Genre and American Culture.* New York: Routledge, 2017.

Brown, Jeffrey A. *Panthers, Hulks, and Ironhearts: Marvel, Diversity, and the 21st Century Superhero.* New Brunswick, NJ: Rutgers University Press, 2021.

Bucciferro, Claudia. "Mutancy, Otherness, and Empathy in the X-Men." In *The X-Men Films: A Cultural Analysis*, edited by Claudia Bucciferro. Lanham, MD: Rowman & Littlefield, 2016.

Buchanan, Bruce G. "A (Very) Brief History of Artificial Intelligence." *AI Magazine* 26, no. 4 (2005): 53–60.

Butler, Judith. *Precarious Life: The Powers of Mourning and Violence*. New York: Verso, 2004.

Butler, Judith. *Frames of War: When is Life Grievable?* London: Verso Books, 2009.

Call, Lewis. "'Alien Commies from the Future!': Diversity, Equity, and Inclusion in Season Seven of *Agents of S.H.I.E.L.D.*" *Slayage* 19, nos. 1–2 (2021): 143–84.

Callicott, J. Baird. "Traditional American Indian and Western European Attitudes Toward Nature: An Overview." *Environmental Ethics* 4 (1982): 293–318.

Calvert, Bronwen. "The AI and the Looking Glass: Embodiment, Virtuality, and Power in *Agents of S.H.I.E.L.D.*, Season 4." *Slayage: The Journal of Whedon Studies* 17, no. 1 (2019): 11–35.

Čapek, Karel, *R.U.R. (Rossum's Universal Robots)*. Translated by Claudia Novack. New York: Penguin Books, 2004.

Carnes, Nicholas, and Lilly J. Green. "An Introduction to the Politics of the Marvel Cinematic Universe." In *The Politics of the Marvel Cinematic Universe*, edited by Nicholas Carnes and Lilly J. Goren. Lawrence: University Press of Kansas, 2023.

Chambliss, Julian C., William L. Svitavsky, and Daniel Fandino, eds. *Assembling the Marvel Cinematic Universe: Essays on the Social, Cultural and Geopolitical Domains*. Jefferson, NC: McFarland & Company, 2018.

Chirot, Daniel, and Clark McCauley. *Why Not Kill Them All? The Logic and Prevention of Mass Political Murder*. Princeton, NJ: Princeton University Press, 2006.

Cho, Sumi, Kimberlé Crenshaw, and Leslie McCall. "Toward a Field of Intersectionality Studies: Theory, Applications, and Praxis." *Signs* 38, no. 4 (2013): 785–810.

Chon-Torres, Octavio. "Disciplinary Nature of Astrobiology and Astrobioethic's Epistemic Foundations." *International Journal of Astrobiology* 20, no. 3 (2021): 186–93.

Chon-Torres, Octavio. "Astrobioethics: Epistemological, Astroethical, and Interplanetary Issues." In *Astrobiology: Science, Ethics, and Public Policy*, edited by Octavio A. Chon-Torres, Ted Peters, Joseph Seckback, and Richard Gordon. Malden, MA: Scrivener, 2021.

Chon-Torres, Octavio, and Konrad Szocik. "A Brief Epistemological Discussion of Astrotheology in the Light of Astrobiology." *International Journal of Astrobiology* 21, no. 1 (2022): 1–8.

Christian, Brian. *The Alignment Problem: Machine Learning and Human Values*. New York: W. W. Norton & Company, 2020.

Chyba, Christopher F., and Kevin P. Hand. "Astrobiology: The Study of the Living Universe." *Annual Review of Astronomy and Astrophysics* 43 (2005): 31–74.

Ćirković, Milan M. "Risks of Space Colonization." *Futures* 126 (2021): 102638.

Ćirković, Milan M. "The Nutshell Kings: Why is Human Space Settlement Controversial in the First Place?" *Futures* 143 (2022): 103023.

Ćirković, Milan M. "Ballet Not For Everyone: A Reply to Szocik." *Futures* 147 (2023): 103116.

Claeys, Gregory. *Dystopia: A Natural History*. Oxford: Oxford University Press, 2017.

Clark, Henry. *The Ethical Mysticism of Albert Schweitzer*. Boston: Beacon Press, 1962.

BIBLIOGRAPHY

Cocca, Carolyn. "Containing the X-Women: De-Powering and De-Queering Female Characters." In *The X-Men Films: A Cultural Analysis*, edited by Claudia Bucciferro. Lanham, MD: Rowman & Littlefield, 2016.

Cocca, Carolyn. *Superwomen: Gender, Power, and Representation.* New York: Bloomsbury Academic, 2016.

Cocca, Carolyn. *Wonder Woman and Captain Marvel: Militarism and Feminism in Comics and Film.* New York: Routledge, 2020.

Cockell, Charles S. "Planetary Protection: A Microbial Ethics Approach." *Space Policy* 21, no. 4 (2005): 287–92.

Cockell, Charles S. "Microbial Rights?" *EMBO Reports* 12, no. 3 (2011): 181.

Coeckelbergh, Mark. *AI Ethics.* Cambridge, MA: The MIT Press, 2020.

Coeckelbergh, Mark. *Robot Ethics.* Cambridge, MA: The MIT Press, 2022.

Collins, Patricia H. *Black Feminist Thought: Knowledge, Consciousness and the Politics of Empowerment.* New York: Routledge, 1991.

Creed, Barbara. *Phallic Panic: Film, Horror and the Primal Uncanny.* Carlton, Australia: Melbourne University Press, 2005.

Crenshaw, Kimberlé. "Mapping the Margins: Intersectionality, Identity Politics, and Violence Against Women of Color." *Stanford Law Review* 43, no. 6 (1991): 1241–99.

Crimston, Charlie R., Paul G. Bain, Matthew J. Hornsey, and Brock Bastian. "Moral Expansiveness: Examining Variability in the Extension of the Moral World." *Journal of Personality and Social Psychology* 111, no. 4 (2016): 636–53.

Crimston, Charlie R., Matthew J. Hornsey, Paul G. Bain, and Brock Bastian. "Toward a Psychology of Moral Expansiveness." *Current Directions in Psychological Science* 27, no. 1 (2018): 14–19.

Crootof, Rebecca. "The Killer Robots are Here: Legal and Policy Implications." *Cardozo Law Review* 36 (2015): 1837–1915.

Curtis, Neal, and Valentina Cardo. "Superheroes and Third-Wave Feminism." *Feminist Media Studies* 18, no. 3 (2018): 381–96.

Daily, Shaundra B., Melva T. James, David Cherry, John J. Porter III, Shelby S. Darnell, Joseph Isaac, and Tania Roy. "Affective Computing: Historical Foundations, Current Applications, and Future Trends." In *Emotions and Affect in Human Factors and Human-Computer Interaction*, edited by Myounghoon Jeon. Cambridge, MA: Academic Press, 2017.

Darowski, Joseph J. *X-Men and the Mutant Metaphor: Race and Gender in the Comic Books.* Lanham, MD: Rowman & Littlefield, 2014.

Darowski, Joseph J. "When Business Improved Art: The 1975 Relaunch of Marvel's Mutant Heroes." In *The Ages of the X-Men: Essays on the Children of the Atom in Changing Times*, edited by Joseph J. Darowski. Jefferson, NC: McFarland & Co., 2014.

Dastur, Françoise. *How Are We to Confront Death? An Introduction to Philosophy.* Translated by Robert Vallier. New York: Fordham University Press, 2012.

Davidson, Joe P. L. "Extinctiopolitics: Existential Risk Studies, the Extintiopolitical Unconscious, and the Billionaire's Exodus from Earth." *New Formations* 107/108 (2022): 48–65.

Davies, Paul. "The Eerie Silence." *Physics World* 23, no. 3 (2010): 28–33.

de Dauw, Esther. *Hot Pants and Spandex Suits: Gender Representation in American Superhero Comic Books*. New Brunswick, NJ: Rutgers University Press, 2021.

de Goede, Marieke. "Fighting the Network: A Critique of the Network as a Security Technology." *Distinktion: Scandinavian Journal of Social Theory* 13, no. 2 (2012): 215–32.

de Goede, Marieke, and Samuel Randalls. "Precaution, Preemption: Arts and Technologies of the Actionable Future." *Environment and Planning D: Society and Space* 27, no. 5 (2009): 859–78.

de Goede, Marieke, and Gavin Sullivan. "The Politics of Security Lists." *Environment and Planning D: Society and Space* 34, no. 1 (2016): 67–88.

Dittmer, Jason. "American Exceptionalism, Visual Effects, and the Post-9/11 Cinematic Superhero Boom." *Environment and Planning D: Society and Space* 29, no. 1 (2011): 114–30.

Dittmer, Jason. "Captain America in the News: Changing Mediascapes and the Appropriation of a Superhero." *Journal of Graphic Novels and Comics* 3, no. 2 (2012): 143–57.

Dixon, Bernard. "Smallpox-Imminent Extinction and an Unresolved Dilemma." *New Scientist* 69, no. 989 (1976): 430–32.

Döbler, Niklas and Marius Raab. "Thinking ET: A Discussion of Exopsychology." *Acta Astronautica* 189 (2021): 699–711.

Domínguez-García, Beatriz. "Of Mutants and Monsters: A Posthuman Study of Verhoeven's and Wiseman's *Total Recall*." *Revista Hélice* 7, no. 1 (2021): 37–51.

Donaldson, Sue, and Will Kymlicka. "The Moral Ark." *Queen's Quarterly* 114, no. 2 (2007): 187–205.

Doncieux, Stephane, Nicholas Bredeche, Jean-Baptiste Mouret, and Agoston E. Eiben. "Evolutionary Robotics: What, Why, and Where To." *Frontiers in Robotics and AI* 2, no. 4 (2015): 1–18.

Douard, John, and Pamela D. Schultz. *Monstrous Crimes and the Failure of Forensic Psychiatry*. New York: Springer, 2013).

Dove-Viebahn, Aviva. "Fantasies of White Feminism: The Human as 'Other' in *Captain Marvel*." *Feminist Media Studies* 23, no. 2 (2023): 1–15.

Drislane, Liam, "The Pretense of Prosthesis: The Prosthecized Superhero in the Marvel Cinematic Universe," *Science Fiction Film and Television* 15, no. 2 (2022): 169–91.

Dyer, Gwynne. *War*. London: Guild Publishing, 1985.

Evans, J. Claude. *With Respect for Nature: Living as Part of the Natural World*. Albany: State University of New York Press, 2018.

Everett, John R. "Albert Schweitzer and Philosophy." *Social Research* 33, no. 4 (1966): 513–30.

Fagundes, Dave. "What We Talk About When We Talk About Persons: The Language of Legal Fiction." *Harvard Law Review* 114, no. 6 (2001): 1745–68.

Fawaz, Ramzi. *The New Mutants: Superheroes and the Radical Imagination of American Comics*. NY: New York University Press, 2016.

Feinberg, Joel. *Rights, Justice, and the Bounds of Liberty: Essays in Social Philosophy*. Princeton, NJ: Princeton University Press, 1980.

Fischer, John M. *Death, Immortality, and Meaning in Life.* New York: Oxford University Press, 2020.

Flanagan, Martin, Mike McKenny, and Amy Livingstone. *The Marvel Studios Phenomenon: Inside a Transmedia Universe.* New York: Bloomsbury, 2016.

Ferrando, Francesca. *Philosophical Posthumanism.* New York: Bloomsbury, 2020.

Floreano, Dario, Francesco Mondada, Andres Perez-Uribe, and Daniel Roggen. "Evolution of Embodied Intelligence." In *Embodied Intelligence: International Seminar, Dagstuhl Castle, Germany, July 7–11, 2003, Revised Papers.* Berlin-Heidelberg: Springer, 2004.

Fogg, Martyn J. "The Ethical Dimensions of Space Settlement." *Space Policy* 16, no. 3 (2000): 205–11.

Foucault, Michel. *The Archaeology of Knowledge and the Discourse on Language.* Translated by A. M. Sheridan Smith. New York: Pantheon Books, 1972.

Foucault, Michel. "About the Concept of the 'Dangerous Individual' in 19th Century Legal Psychiatry." *International Journal of Law and Psychiatry* 1, no. 1 (1978): 1–18.

François, Agathe. "The Figuration of Post-Human Bodies: A Processual Experiment with Imaginaries." *Im@go: A Journal of the Social Imaginary* 12, no. 7 (2018): 25–38.

Frankl, Viktor E. *Man's Search for Meaning.* Boston: Beacon Press, 2006.

Friedman, Lester D., and Allison B. Kavey. *Monstrous Progeny: A History of the Frankenstein Narratives.* New Brunswick, NJ: Rutgers University Press, 2016.

Frierson, Patrick R. *What Is the Human Being?* New York: Routledge, 2013.

Fröding, Barbro, and Martin Peterson. "Friendly AI." *Ethics and Information Technology* 23, no. 3 (2021): 207–14.

Galliot, Jai, and Mianna Lotz, eds. *Super Soldiers: The Ethical, Legal, and Social Implications.* Aldershot, UK: Ashgate, 2015.

Galway-Witham, Julie, and Chris Stringer. "How did *Homo sapiens* Evolve?" *Science* 360, no. 6395 (2018): 1296–98.

Garneau, Eric, and Maura Foley. "Grant Morrison's Mutants and the Post-9/11 Culture of Fear." In *The Ages of the X-Men: Essays on the Children of the Atom in Changing Times,* edited by Joseph J. Darowski. Jefferson, NC: McFarland & Co., 2014.

Garrido, Daniel R. "Deaths of the Subject and Negated Subjectivity in the Era of Neoliberal Capitalism." *tripleC* 17, no. 1 (2019): 159–84.

Gilligan, James. *Violence: Reflections on a National Epidemic.* New York: Vintage Books, 1997.

Gilroy, Paul. *Against Race; Imagining Political Culture Beyond the Color Line.* Cambridge, MA: Harvard University Press, 2000.

Gittinger, Juli. *Personhood in Science Fiction: Religious and Philosophical Considerations.* Cham, Switzerland: Palgrave Macmillan, 2019.

Glenn, Linda M. "What is a Person?" In *Posthumanism: The Future of Homo Sapiens,* edited by Michael Bess and Diana Walsh Pasulka. Farmington Hills, MI: Macmillan, 2018.

Globokar, Roman. "The Ethics of Albert Schweitzer as an Inspiration for Global Ethics." *Studia Teologiczno-Historyczne* 40, no. 1 (2020): 55–71.

Goertzel, Ben. "Human-Level Artificial General Intelligence and the Possibility of a Technological Singularity: A Reaction to Ray Kurzweil's *The Singularity is Near,* and McDermott's Critique of Kurzweil." *Artificial Intelligence* 171, no. 18 (2007): 1161–73.

Goertzel, Ben. "Artificial General Intelligence: Concept, State of the Art, and Future Prospects." *Journal of Artificial General Intelligence* 5, no. 1 (2014): 1–46.

Gomel, Elana. *Representations of the Post/Human: Monsters, Aliens and Others in Popular Culture.* New Brunswick, NJ: Rutgers University Press, 2002.

Gomel, Elana. "Posthuman Rights: The Ethics of Alien Encounter." In *Unveiling the Posthuman*, edited by Artur Matos Alves. Oxford: Inter-Disciplinary Press, 2012.

Gomel, Elana. *Science Fiction, Alien Encounters, and the Ethics of Posthumanism: Beyond the Golden Rule.* New York: Palgrave Macmillan, 2014.

Good, Irving J. "Speculations Concerning the First Ultraintelligent Machine." *Advances in Computers* 6 (1965): 31–88.

Gould, Stephen J. "The Evolution of Life on the Earth." *Scientific American* 271, no. 4 (1994): 84–91.

Green, Stephanie. "Playing at Being a Superhero: Trish Walker in Jessica Jones." *Imagining the Impossible: International Journal for the Fantastic in Contemporary Media* 1, no. 1 (2022): 1–16.

Grossman, Dave. *On Killing: The Psychological Cost of Learning to Kill in War and Society.* New York: Back Bay Books, 1996.

Gunderson, Ryan, Diana Stuart, and Brian Petersen. "In Search of Plan(et) B: Irrational Rationality, Capitalist Realism, and Space Colonization." *Futures* 134 (2021): 102857.

Gunkle, David J. *The Machine Question: Critical Perspectives on AI, Robots, and Ethics.* Cambridge, MA: The MIT Press, 2017.

Gunkle, David J. *How to Survive a Robot Invasion: Rights, Responsibility, and AI.* New York: Routledge, 2020.

Gurkaynak, Gonenc, Ilay Yilmaz, and Gunes Haksever. "Stifling Artificial Intelligence: Human Perils." *Computer Law & Security Review* 32, no. 5 (2016): 749–58.

Hacking, Ian. "Making Up People." In *Reconstructing Individualism: Autonomy, Individuality, and the Self in Western Thought*, edited by Thomas C. Heller, Morton Sosna, and David E. Wellbery. Palo Alto, CA: Stanford University Press, 1986.

Harrison Dinnis, Heather A., and Jann K. Kleffner. "Solider 2.0: Military Human Enhancement and International Law." *International Law Studies* 92, no. 1 (2016): 432–82.

Haenlein, Michael, and Andreas Kaplan. "A Brief History of Artificial Intelligence: On the Past, Present, and Future of Artificial Intelligence." *California Management Review* 61, no. 4 (2019): 5–14.

Hägglund, Martin. *This Life: Why Mortality Makes Us Free.* London: Profile Books, 2019.

Halberstam, Judith. *Skin Shows: Gothic Horror and the Technology of Monsters.* Durham, NC: Duke University Press, 1995.

Hall, John Storrs. *Beyond AI: Creating the Conscience of the Machine.* New York: Prometheus Books, 2007.

Hart, Emma, and Léni K. Le Goff. "Artificial Evolution of Robot Bodies and Control: On the Interaction Between Evolution, Learning and Culture." *Philosophical Transactions of the Royal Society B* 377, no. 1843 (2022): 1–8.

Haslem, Wender, Elizabeth MacFarlane, and Sarah Richardson, eds. *Superhero Bodies: Identity, Materiality, Transformation.* New York: Routledge, 2018.

Hazen, Robert. "What is Life?" *New Scientist* 192, no. 2578 (2006): 46–51.

Hebenton, Bill, and Toby Seddon. "From Dangerousness to Precaution: Managing Sexual and Violent Offenders in an Insecure and Uncertain Age." *The British Journal of Criminology* 49, no. 3 (2009): 343–62.

Herbrechter, Stefan. *Posthumanism: A Critical Analysis*. New York: Bloomsbury, 2013.

Holmberg, Christine, Christine Bischof, and Susanne Bauer. "Making Predictions: Computing Populations." *Science, Technolgy, & Human Values* 38, no. 3 (2013): 398–420.

Hull, Matthew S. "Documents and Bureaucracy." *Annual Review of Anthropology* 41, no. 1 (2012): 251–67.

Irwin, Louis, and Dirk Schulze-Makuch. "The Astrobiology of Alien Worlds: Known and Unknown Forms of Life." *Universe* 6, no. 9 (2020): 1-32.

Jackson, Zakiyyah I. "Animal: New Directions in the Theorization of Race and Posthumanism." *Feminist Studies* 39, no. 3 (2013): 669–85.

Jackson, Zakiyyah I. "Outer Worlds: The Persistence of Race in Movement 'Beyond the Human.'" *GLQ: A Journal of Lesbian and Gay Studies* 21, no. 2 (2015): 215–18.

Jacob, Marie-Andrée. "Form-Made Persons: Consent Forms as Consent's Blind Spot." *PoLAR: Political and Legal Anthropology Review* 30, no. 2 (2007): 249–68.

Jaynes, Tyler L. "On Human Genome Manipulation and *Homo technicus*: The Legal Treatment of Non-Natural Human Subjects." *AI and Ethics* 1, no. 3 (2021): 331–45.

Joffe, Robyn. "Holding Out for a Hero(ine): An Examination of the Presentation and Treatment of Female Superheroes in Marvel Movies." *Panic at the Disorder: An Interdisciplinary Journal* 1, no. 1 (2019): 5–19.

Jonas, Hans. *The Imperative of Responsibility: In Search of an Ethics for the Technological Age*. Chicago: University of Chicago Press, 1984.

Jordan, John M. *Robots*. Cambridge, MA: The MIT Press, 2016.

Katz, Eric. "The Liberation of Humanity and Nature." *Environmental Values* 11, no. 4 (2002): 397–405.

Keiper, Adam, and Ari N. Schulman. "The Problem with 'Friendly' Artificial Intelligence." *The New Atlantis* 32 (2011): 80–89.

Kellehear, Allan. *A Social History of Dying*. New York: Cambridge University Press, 2007.

Kent, Miriam. "'Let's Rewrite Some History, Shall We?': Temporality and Postfeminism in *Captain Marvel*'s Comic Book Superhero(ine)ism." *Feminist Media Studies* 23, no. 2 (2023): 394–410.

Kerman, Judith, ed. *Retrofitting Blade Runner: Issues in Ridley Scott's Blade Runner and Philip K. Dick's Do Androids Dream of Electric Sheep?* Madison: University of Wisconsin Press, 1997.

King, Geoff, and Tanya Krzywinska. *Science Fiction Cinema: From Outerspace to Cyberspace*. London: Wallflower Press, 2000.

Kingwell, Mark. *Singular Creatures: Robots, Rights, and the Politics of Posthumanism*. Montreal & Kingston: McGill-Queen's University Press, 2022.

Kitchin, Rob, and James Kneale, eds. *Lost in Space: Geographies of Science Fiction*. New York: Continuum, 2002.

Kline, Ronald R. "Cybernetics, Automata Studies, and the Dartmouth Conference on Artificial Intelligence." *IEEE Annals of the History of Computing* 33, no. 4 (2010): 5–16.

Kountz, Samantha. "We Come in Peace: Immigration in Post-Cold War Science Fiction Film." *Foundation* 43, no. 119 (2014): 29–40.

Kripal, Jeffrey J. *Mutants and Mystics: Science Fiction, Superhero Comics, and the Paranormal.* Chicago: University of Chicago Press, 2011.

Küchler, Ulrike, Silja Maehl, and Graeme Stout, eds. *Alien Imaginations: Science Fiction and Tales of Transnationalism.* New York: Bloomsbury, 2015.

Kurzweil, Ray. *The Singularity is Near: When Humans Transcend Biology.* New York: Penguin, 2006.

Kwok, Sun. *Stardust: The Cosmic Seeds of Life.* Berlin: Springer, 2013.

Lakoff, George. "Introduction." In *Louder than Words: The New Science of How the Mind Makes Meaning,* edited by Benjamin K. Bergen. New York: Basic Books, 2012.

Larsen, Soren C., and Jay T. Johnson. "In Between Worlds: Place, Experience, and Research in Indigenous Geography." *Journal of Cultural Geography* 29, no. 1 (2012): 1–13.

Leander, Anna. "The Politics of Whitelisting: Regulatory Work and Topologies in Commercial Security." *Environment and Planning D: Society and Space* 34, no. 1 (2016): 48–66.

Lechuga, Michael. *Visions of Invasion: Alien Affects, Cinema, and Citizenship in Settler Colonies.* Jackson: University Press of Mississippi, 2023.

le Clue, Natalie, and Janelle Vermaak-Griessel. "Artificial Feminism: Fan Reaction to the Representation of Captain Marvel." In *Gender and Action Films: 2000 and Beyond,* edited by Steven Gerrard and Renee Middlemost. Leeds, UK: Emerald Publishing, 2022.

Lederberg, Joshua. "Exobiology: Approaches to Life Beyond the Earth." *Science* 132, no. 3424 (1960): 393–400.

Lee, Andrew Y. "Speciesism and Sentientism." *Journal of Consciousness Studies* 29, nos. 3–4 (2022): 205–28.

Lee, Edward. "Are We Losing Control?" In *Perspectives on Digital Humanism,* edited by Hannes Werthner, Erich Prem, Edward A. Lee, and Carlo Ghezzi. Cham, Switzerland: Springer, 2022.

Leitenberg, Milton. "Deaths in Wars and Conflicts in the 20th Century." *Cornell University Peace Studies Program Occasional Paper #29.* New Haven, CT: Cornell University Press, 2006.

Levina, Marina, and Diem-My T. Bui. "Introduction: Toward a Comprehensive Monster Theory in the 21st Century." In *Monster Culture in the 21st Century: A Reader,* edited by Marina Levina and Diem-My T. Bui. London: Bloomsbury, 2013.

Li, Oliver. "Problems with 'Friendly AI.'" *Ethics and Information Technology* 23, no. 3 (2021): 543–50.

Liivoja, Rain. "Being More Than You Can Be: Enhancement of Warfighters and the Law of Armed Conflict." *Law and the Future of War Research Paper No. 1.* Queensland, Australia: University of Queensland Law School, 2020.

Lin, Patrick, Keith Abney, and George A. Bekey, eds. *Robot Ethics: The Ethical and Social Implications of Robotics.* Cambridge, MA: The MIT Press, 2008.

Lizza, John P. *Persons, Humanity, and the Definition of Death.* Baltimore: Johns Hopkins University Press, 2006.

BIBLIOGRAPHY

Lord, Christina. "Facing the Science-Fictional Other: Human-Alien Contact in J. H Rosney Aîné's *Les Xipéhuz*." *Nineteenth-Century French Studies* 48, nos. 1–2 (2019–2020): 149–65.

Lucanio, Patrick. *Them or Us: Archetypal Interpretations of Fifties Alien Invasion Films.* Bloomington: Indiana University Press, 1987.

Luper, Steven. *The Philosophy of Death.* Cambridge: Cambridge University Press, 2009.

Lupisella, Mark. "Astrobiology and Cosmocentrism." *Bioastronomy News* 10, no. 1 (1998): 1–2, 8.

Lyons, Siobhan. *Death and the Machine: Intersections of Mortality and Robotics.* Singapore: Palgrave Macmillan, 2018.

MacCormack, Patricia. "Queer Posthumanism: Cyborgs, Animals, Monsters, Perverts." In *The Ashgate Research Companion to Queer Theory*, edited by Noreen Giffney and Michael O'Rourke. New York: Routledge, 2009.

Maddrell, Avril, and James D. Sidaway, eds. *Deathscapes: Spaces for Death, Dying, Mourning and Remembrance.* Aldershot, UK: Ashgate, 2012.

Marchant, Gary E., Braden Allenby, Ronald Arkin, Edward T. Barrett, Jason Borenstein, Lyn M. Gaudet, Orde Kitrie, Patrick Lin, George R. Lucas, Richard O'Meara, and Jared Silberman. "International Governance of Autonomous Military Robots." *The Columbia Science and Technology Law Review* 12 (2011): 272–315.

Marino, Allesandro. "Astroenvironmentalism as SF: Bordering (and Ordering) Otherworldly Ecologies." *Environmental Humanities* 15, no. 1 (2023): 25–43.

Marino, Alessandra, and Thomas Cheney. "Centering Environmentalism in Space Governance: Interrogating Dominance and Authority Through a Critical Legal Geography of Outer Space." *Space Policy* 63 (2023): 1–10.

Marino, Lori. "Humanity is Not Prepared to Colonize Mars." *Futures* 110 (2019): 15–18.

Markowitz, Judith A. *Robots that Kill: Deadly Machines and their Precursors in Myth, Folklore, Literature, Popular Culture and Reality.* Jefferson, NC: McFarland, 2019.

Marshall, Alan. "Development and Imperialism in Space." *Space Policy* 11, no. 1 (1995): 41–52.

Marshall, Catherine. "Refugees and Other Aliens." *Eureka Street* 19, no. 16 (2009): 38–39.

Marshall, S. L. A. *Men Against Fire.* Glouchester, UK: Peter Smith, 1978.

Martin, Mike W. *Albert Schweitzer's Reverence for Life: Ethical Idealism and Self-Realization.* Aldershot, UK: Ashgate, 2012.

May, Todd, *Death.* Durham, UK: Acumen, 2014.

May, Todd, *A Significant Life: Human Meaning in a Silent Universe.* Chicago: University of Chicago Press, 2015.

Mays, James L. "What is a Human Being? Reflections on Psalm 8." *Theology Today* 50, no. 4 (1994): 511–20.

McCarthy, John, Marvin L. Minsky, Nathaniel Rochester, and Claude E. Shannon. "A Proposal for the Dartmouth Summer Research Project on Artificial Intelligence, August 31, 1955." *AI Magazine* 27, no. 4 (2006): 12–14.

McCormack, Donna. "Monster Talk: A Virtual Roundtable with Mark Bould, Liv Bugge, Surekha Davies, Margrit Shildrick, and Jeffery Weinstock." *Somatechnics* 8, no. 2 (2018): 248–68.

McDannell, Colleen, and Bernhard Lang. *Heaven: A History*. New Haven, CT: Yale University Press, 2001.

McMahan, Jeff. *The Ethics of Killing*. Oxford: Oxford University Press, 2002.

McMahan, Jeff. *Killing by Remote Control: The Ethics of an Unmanned Military*. Oxford: Oxford University Press, 2013.

McSweeney, Terence. *Avengers Assemble! Critical Perspectives on the Marvel Cinematic Universe*. New York: Columbia University Press, 2018.

McSweeney, Terence. *The Contemporary Superhero Film: Projections of Power and Identity*. New York: Columbia University Press, 2020.

Medina-Contreras, Juan, and Pedro Sangro Colón. "Representations of Defense Organizations in the Marvel Cinematic Universe (2008–2019)." *Communication & Society* 33, no. 4 (2020): 19–32.

Meehan, Katherine, Ian G. R. Shaw, and Sallie Marston. "Political Geographies of the Object." *Political Geography* 33 (2013): 1–10.

Meltzer, Michael. *When Biospheres Collide: A History of NASA's Planetary Protection Programs*. Washington, DC: US Government Printing Office, 2010.

Mesler, Bill, and H. James Cleaves II. *A Brief History of Creation: Science and the Search for the Origin of Life*. New York: W. W. Norton & Company, 2016.

Metz, Thaddeus. *Meaning in Life: An Analytic Study*. Oxford: Oxford University Press, 2013.

Metzl, Jamie. *Hacking Darwin: Genetic Engineering and the Future of Humanity*. Naperville, IL: Sourcebooks, 2019.

Meyer, Marvin. "Affirming Reverence for Life." In *Reverence for Life: The Ethics of Albert Schweitzer for the Twenty-First Century*, edited by Marvin Meyer and Kurt Bergel. Syracuse, NY: Syracuse University Press, 2002.

Miller, P. Andrew. "Mutants, Metaphor, and Marginalism: What X-actly Do the X-Men Stand For?" *Journal of the Fantastic in the Arts* 13, no. 3 (2003): 282–90.

Morgan, Jamie. "Species Being in the Twenty-First Century." *Review of Political Economy* 30, no. 3 (2018): 377–95.

Mori, Masahiro. "The Uncanny Valley." *IEEE Robotics & Automation Magazine* 19, no. 2 (2012): 98–100.

Moynihan, Thomas. *X-Risk: How Humanity Discovered its Own Extinction*. Falmouth, UK: Urbanomic, 2020.

Muehlhauser, Luke, and Nick Bostrom. "Why We Need Friendly AI." *Think* 36, no. 13 (2014): 41–47.

Muller, Michael. "Multiple Paradigms in Affective Computing." *Interacting with Computers* 16, no. 4 (2004): 759–68.

Murphy, Robin R. "The Original 'I, Robot,' Featured a Murderous Robot and the Frankenstein Complex." *Science Robotics* 7, no. 71 (2022): 1-2.

Murray, Mitch. "The Work of Art in the Age of the Superhero." *Science Fiction Film and Television* 10, no. 1 (2017): 27–51.

Musk, Elon. "Making Humans a Multi-Planetary Species." *New Space* 5, no. 2 (2017): 46–61.

Nadkarni, Samira Shirish. "'To Be the Shield': American Imperialism and Explosive Identity Politics in *Agents of S.H.I.E.L.D.*" In *Assembling the Marvel Cinematic Universe:*

Essays on the Social, Cultural and Geopolitical Domains, edited by Julian C. Chambliss, William L. Svitavsky, and Daniel Fandino. Jefferson, NC: McFarland & Co., 2018.

Nash, Roderick F. *The Rights of Nature: A History of Environmental Ethics*. Madison: University of Wisconsin Press, 1989.

Nath, Rajakishore, and Vineet Sahu. "The Problem of Machine Ethics in Artificial Intelligence." *AI & Society* 35 (2020): 103–11.

Nayar, Pramod. *Posthumanism*. Malden, MA: Polity Press, 2014.

Nee, Sean. "The Great Chain of Being." *Nature* 435, no. 7041 (2005): 429.

Newman, Kim. "Mutants and Monsters." In *It Came from the 1950s! Popular Culture, Popular Anxieties*, edited by Darryl Jones, Elizabeth McCarthy, and Bernice M. Murphy. New York: Palgrave Macmillan, 2011.

Nurse, Paul. *What is Life? Five Great Ideas in Biology*. New York: W. W. Norton & Company, 2021.

Nussbaum, Martha C. *The Therapy of Desire: Theory and Practice in Hellenistic Ethics*. Princeton, NJ: Princeton University Press, 2013.

Nyholm, Sven. *Humans and Robots: Ethics, Agency, and Anthropomorphism*. New York: Rowman & Littlefield, 2020.

Nyholm, Sven. "A New Control Problem? Humanoid Robots, Artificial Intelligence, and the Value of Control." *AI and Ethics* 3, no. 4 (2023): 1229-39.

Oermann, Nils O. *Albert Schweitzer: A Biography*. Oxford: Oxford University Press, 2016.

Oravec, Jo Ann. *Good Robot, Bad Robot: Dark and Creepy Sides of Robotics, Autonomous Vehicles, and AI*. Cham, Switzerland: Springer Nature, 2022.

Ord, Toby. *The Precipice: Existential Risk and the Future of Humanity*. New York: Hachette Books, 2020.

Paget, James C., and Michael J. Thate, eds. *Albert Schweitzer in Thought and Action: A Life in Parts*. Syracuse, NY: Syracuse University Press, 2016.

Pardy, Brett. "The Militarization of Marvel's Avengers." *Studies in Popular Culture* 42, no.1 (2019): 103–22.

Parks, Gregory S., and Matthew W. Hughey. "'A Choice of Weapons': The X-Men and the Metaphor for Approaches to Racial Equality." *Indiana Law Journal* 92, no. 5 (2017): 1–26.

Pepperell, Robert. *The Post-Human Condition*. Bristol, UK: Intellect Books, 2003.

Picard, Rosalind W. *Affective Computing*. Cambridge, MA: The MIT Press, 1997.

Picard, Rosalind W. "Response to Sloman's Review of *Affective Computing*." *AI Magazine* 20, no. 1 (1999): 134–37.

Picard, Rosalind W. "Affective Computing: Challenges." *International Journal of Human-Computer Studies* 59, nos. 1–2 (2003): 55–64.

Pilisuk, Marc, with Jennifer A. Rountree. *Who Benefits from Global Violence and War: Uncovering a Destructive System*. Westport, CT: Praeger Security, 2008.

Porpora, Douglas V. "Dehumanization in Theory: Anti-Humanism, Non-Humanism, Post-Humanism, and Trans-Humanism." *Journal of Critical Realism* 16, no. 4 (2017): 353–67.

Pratt, John. "Governing the Dangerous: An Historical Overview of Dangerous Offender Legislation." *Social & Legal Studies* 5, no. 1 (1996): 21–26.

Ratto, Casey M. "Not Superhero Accessible: The Temporal Stickiness of Disability in Superhero Comics." *Disability Studies Quarterly* 37, no. 2 (2017): n.p.

Rees, Martin J. *Our Final Hour: A Scientist's Warning: How Terror, Error, and Environmental Disaster Threaten Humankind's Future in this Century—On Earth and Beyond*. New York: Basic Books, 2003.

Ricard, Matthieu. *A Plea for the Animals: The Moral, Philosophical, and Evolutionary Imperative to Treat All Beings with Compassion*. Boulder, CO: Shambhala Publications, 2016.

Rieder, John. "Science Fiction, Colonialism, and the Plot of Invasion." *Extrapolation* 46, no. 3 (2005): 373–94.

Rieder, John. *Colonialism and the Emergence of Science Fiction*. Middletown, CT: Wesleyan University Press, 2008.

Robbins, Paul, John Hintz, and Sarah A. Moore. *Environment and Society: A Critical Introduction*. New York: Wiley Blackwell, 2014.

Robinson, Ashley. "We Are Iron Man: Tony Stark, Iron Man, and American Identity in the Marvel Cinematic Universe's Phase One Films." *The Journal of Popular Culture* 51, no. 4 (2018): 824–44.

Robinson, Joanna, Dave Gonzales, and Gavin Edwards. *MCU: The Reign of Marvel Studios*. New York: Liveright Publishing, 2024.

Rodman, John. "The Liberation of Nature?" *Inquiry* 20, nos. 1–4 (1977): 83–131.

Rodogno, Raffaele. "Sentientism, Wellbeing, and Environmentalism." *Journal of Applied Philosophy* 27, no. 1 (2010): 84–99.

Roitblat, Herbert L. *Algorithms are Not Enough: Creating General Artificial Intelligence*. Cambridge, MA: The MIT Press, 2020.

Roli, Andrea, Johannes Jaeger, and Stuart Kauffman. "How Organisms Come to Know the World: Fundamental Limits on Artificial General Intelligence." *Frontiers in Ecology and Evolution* 9 (2022): 1–14.

Romagnoli, Alex S., and Gian S. Pagnucci. *Enter the Superheroes: American Values, Culture, and the Canon of Superhero Literature*. Lanham, MD: Scarecrow Press, 2013.

Ruin, Hans. *Being with the Death: Burial, Ancestral Politics, and the Roots of Historical Consciousness*. Palo Alto, CA: Stanford University Press, 2018.

Ruíz, Elena. "Framing Intersectionality." In *The Routledge Companion to Philosophy of Race*, edited by Paul C. Taylor, Linda Martín Alcoff, and Luvell Anderson. New York: Routledge, 2017.

Russell, Stuart. *Human Compatible: Artificial Intelligence and the Problem of Control*. New York: Penguin Books, 2020.

Sagan, Carl. *Cosmos*. London: Abacus, 2003.

Saputri, Theresia R. D., and Seok-Won Lee. "The Application of Machine Learning in Self-Adaptive Systems: A Systematic Literature Review." *IEEE Access* 8 (2020): 205948–205967.

Scheffler, Samuel. *Death and the Afterlife*. New York: Oxford University Press, 2016.

Scheu, Johannes. "Dangerous Classes: Tracing Back and Epistemological Fear." *Distinktion: Scandinavia Journal of Social Theory* 12, no. 2 (2011): 115–34.

Schönfeld, Martin. "Who or What has Moral Standing?" *American Philosophical Quarterly* 29, no. 4 (1992): 353–62.

Schrödinger, Erwin. *What is Life? The Physical Aspect of the Living Cell and Mind*. Cambridge: Cambridge University Press, 1944.

BIBLIOGRAPHY

Schulze-Makuch, Dirk, and Paul Davies. "Destination Mars: Colonization via Initial One-Way Missions." *Journal of the British Interplanetary Society* 66 (2013): 11–14.

Schumacher, Bernard M. *Death and Mortality in Contemporary Philosophy*. Translated by Michael J. Miller. Cambridge: Cambridge University Press, 2011.

Schwartz, James S. J. "On the Moral Permissibility of Terraforming." *Ethics and the Environment* 18, no.2 (2013): 1–31.

Schwartz, James S. J., Sheri Wells-Jensen, John W. Traphagan, Deana Weibel, and Kelly Smith. "What Do We Need to Ask Before Settling Space?" *Journal of the British Interplanetary Society* 74 (2021): 2–9.

Schweitzer, Albert. *The Philosophy of Civilization*. Translated by C. T. Campion. Amherst, NY: Prometheus Books, 1987.

Seaver, George. *Albert Schweitzer: The Man and His Mind*. London: A&C Black, 1947.

Segal, Alan. *Life After Death: A History of the Afterlife in Western Religion*. New York: Doubleday, 2010.

Sen, Amartya. *Identity and Violence: The Illusion of Destiny*. New York: W. W. Norton & Company, 2006.

Shalit, Benjamin. *The Psychology of Conflict and Combat*. New York: Prager, 1988.

Shanahan, Murray. *The Technological Singularity*. Cambridge, MA: The MIT Press, 2015.

Shanahan, Timothy. *Philosophy and Blade Runner*. New York: Palgrave Macmillan, 2014.

Shaver, Peter. *Cosmic Heritage: Evolution from the Big Bang to Conscious Life*. London: Springer, 2011.

Shyminsky, Neil. "Mutant Readers, Reading Mutants: Appropriation, Assimilation, and the X-Men." *International Journal of Comic Arts* 8, no. 2 (2006): 387–405.

Silva, Fernando, Miguel Duarte, Luís Correia, Sancho Moura Oliveira, and Anders Lyhne Christensen. "Open Issues in Evolutionary Robotics." *Evolutionary Computation* 24, no. 2 (2016): 205–36.

Sibley, David. *Geographies of Exclusion: Society and Difference in the West*. London: Routledge, 1995.

Singer, Peter. *Animal Liberation: A New Ethics for Our Treatment of Animals*. New York: New York Review, 1975.

Singer, Peter. *Practical Ethics*. New York: Cambridge University Press, 1993.

Singer, Peter. *Animal Liberation Now: The Definitive Classic Renewed*. New York: Harper Perennial, 2015.

Singer, Peter W. *Wired for War: The Robotics Revolution and Conflict in the 21st Century*. New York: Penguin, 2009.

Slijepčevič, Pregrag B. *Biocivilizations: A New Look at the Science of Life*. White River Junction, VT: Chelsea Green, 2023.

Sloman, Aaron. "Review of *Affective Computing*." *AI Magazine* 20, no. 1 (1999): 127.

Smiles, Niiyokamigaabaw Deondre. "Reflections on the (Continued and Future) Importance of Indigenous Geographies." *Dialogues in Human Geography* 14, no. 2 (2024): 217-20.

Smith, Kelly, and Keith Abney. "Human Colonization: A World Too Far?" *Futures* 110 (2019): 1–3.

Soyka, David. "Frankenstein and the Miltonic Creation of Evil." *Extrapolation* 33, no. 2 (1992): 166–77.

Spade, Dean. *Normal Life: Administrative Violence, Critical Trans Politics, and the Limits of Law*. Brooklyn, NY: South End Press, 2011.

Sparrow, Robert. "The Turing Triage Test." *Ethics and Information Technology* 6 (2004): 203–13.

Sparrow, Robert. "Friendly AI will Still be Our Master. Or, Why We Should Not Want to be the Pets of Super-Intelligent Computers." *AI & Society* 39, no. 5 (2024): 2439–44.

Stäheli, Urs. "Indexing—the Politics of Invisibility." *Environment and Planning D: Society and Space* 34, no. 1 (2016): 14–29.

Stapledon, Olaf. *Odd John: A Story Between Jest and Earnest*. London: Methuen, 1935.

Starbuck, William H. "Shouldn't Organization Theory Emerge from Adolescence?" *Organization* 10, no. 3 (2003): 439–52.

Steels, Brian. "Forever Guilty: Convict Perceptions of Pre and Post Conviction." *Current Issues in Criminal Justice* 21, no. 2 (2009): 242–56.

Steinbock, Daniel J. "Designating the Dangerous: From Blacklists to Watch Lists." *Seattle University Law Review* 30 (2006): 65–118.

Stevenson, Olivia, Charlotte Kenten, and Avril Maddrell. "And Now the End is Near: Enlivening and Politicizing the Geographies of Dying, Death and Mourning." *Social & Cultural Geography* 17, no. 2 (2016): 153–65.

Stone, Christopher D. "Should Trees have Standing?—Toward Legal Rights for Natural Objects." *Southern California Law Review* 45 (1972): 450–501.

Street, Alice. "Seen by the State: Bureaucracy, Visibility and Governmentality in a Papua New Guinean Hospital." *The Australian Journal of Anthropology* 23, no. 1 (2012): 1–21.

Szocik, Konrad. "Should and Could Humans Go to Mars? Yes, But Not Now and Not in the Near Future." *Futures* 105 (2019): 54–66.

Szocik, Konrad. "Space Not For Everyone: The Problem of Social Exclusion in the Concept of Space Settlement." *Futures* 145 (2023): 103073.

Szollosy, Michael. "Freud, Frankenstein and our Fear of Robots: Projection in our Cultural Perception of Technology." *AI & Society* 32 (2017): 433–39.

Tajalli, Payman. "AI Ethics and the Banality of Evil." *Ethics and Information Technology* 23, no. 3 (2021): 447–54.

Tamm, Marek, and Zoltán Boldizsár Simon. "Historical Thinking and the Human: Introduction." *Journal of the Philosophy of History* 14, no. 3 (2020): 285–309.

Taylor, Affrica. "Romancing or Re-Configuring Nature in the Anthropocene? Towards Common Worlding Pedagogies." In *Reimagining Sustainability in Precarious Times*, edited by Karen Malone, Son Truong, and Tonia Gray. Singapore: Springer Nature, 2017.

Taylor, Jessica, and Laura Glitsos, "'Having it Both Ways': Containing the Champions of Feminism in Female-Led Origin and Solo Superhero Films." *Feminist Media Studies* 23, no. 2 (2023): 656–70.

Taylor, Kathleen. *Cruelty: Human Evil and the Human Brain*. Oxford: Oxford University Press, 2009.

Taylor, Noah B. *Existential Risks in Peace and Conflict Studies*. Cham, Switzerland: Springer Nature, 2023.

Tegmark, Max. *Life 3.0: Being Human in the Age of Artificial Intelligence*. New York: Vintage, 2018.

BIBLIOGRAPHY

Torres, Émile P. *Human Extinction: A History of the Science and Ethics of Annihilation.* New York: Routledge, 2024.

Traphagan, John W. "Which Humanity Would Space Colonization Save?" *Futures* 110 (2019): 47–49.

Trushell, John M. "American Dreams of Mutants: The X-Men—Pulp Fiction, Science Fiction, and Superheros." *The Journal of Popular Culture* 38, no. 1 (2004): 149–68.

Tucker, Reed. *Slugfest: Inside the Epic 50-Year Battle Between Marvel and DC.* London: Sphere, 2017.

Turner, Nancy J., and Andrea J. Reid. "'When the Wild Roses Bloom': Indigenous Knowledge and Environmental Change in Northwestern North America." *GeoHealth* 6, no. 11 (2022): 1-21.

Underwood, R.S. "Are We Alone in the Universe?" *The Scientific Monthly* 49, no. 2 (1939): 155–59.

Veldhuizen, Vera "Classifying Monsters." In *Interrogating Boundaries of the Nonhuman: Literature, Climate Change, and Environmental Crises,* edited by Matthias Stephan and Sune Borkfelt. Lanham, MD: Lexington Press, 2022.

Veruggio, Gianmarco, and Fiorella Operto. "Roboethics: A Bottom-Up Interdisciplinary Discourse in the Field of Applied Ethics in Robotics." *International Review of Information Ethics* 6, no. 12 (2006): 2–8.

Vint, Sherryl. *Animal Alterity: Science Fiction and the Question of the Anima.* Liverpool, UK: Liverpool University Press, 2010.

Vint, Sherryl. *Science Fiction.* Cambridge, MA: The MIT Press, 2021

Vismann, Cornelia. *Files: Law and Media.* Translated by Geoffrey Winthrop-Young. Palo Alto, CA: Stanford University Press, 2008.

Walker, Paul, and Terence Lovat. "Concepts of Personhood and Autonomy as They Apply to End-of-Life Decisions in Intensive Care." *Medical Health Care and Philosophy* 18 (2015): 309–15.

Waller, James. *Becoming Evil: How Ordinary People Commit Genocide and Mass Killing.* New York: Oxford University Press, 2002.

Wang, Haoyang, and Christina Zhang. "Marvel Cinematic Universe Villains and Social Anxieties." In *The Politics of the Marvel Cinematic Universe,* edited by Nicholas Carnes and Lilly J. Goren. Lawrence: University Press of Kansas, 2023.

Wang, Pei. "On Defining Artificial Intelligence." *Journal of Artificial General Intelligence* 10, no. 2 (2019): 1–37.

Watson, James D. E. "The Harm of Premature Death: Immortality—the Transhumanist Challenge." *Ethical Perspectives* 16, no. 4 (2009): 435–58.

Weitzenfeld, Adam, and Melanie Joy. "An Overview of Anthropocentrism, Humanism, and Speciesism in Critical Animal Theory." *Counterpoints* 448 (2014): 3–27.

Weldes, Jutta, ed. *To Seek Out New Worlds: Science Fiction and World Politics.* New York: Palgrave Macmillan, 2003.

Westfahl, Gary, ed. *Space and Beyond: The Frontier Theme in Science Fiction.* Westport, CT: Greenwood, 2000.

White, Lynn, Jr. "The Historical Roots of Our Ecological Crisis." *Science* 155, no. 3767 (1967): 1203–7.

Williams, Lynda. "Irrational Dreams of Space Colonization." *Peace Review* 22, no. 1 (2010): 4–8.

Wilson, Daniel J. "Lovejoy's the Great Chain of Being After Fifty Years." *Journal of the History of Ideas* 48, no. 2 (1987): 187–206.

Winch, Peter. "Presidential Address: 'Eine Einstellung zur Seele.'" *Proceedings of the Aristotelian Society* 81 (1980–1981): 1–15.

Wolf, Susan. *Meaning in Life and Why it Matters*. Princeton, NJ: Princeton University Press, 2010.

Wynter, Silvia. "No Humans Involved: An Open Letter to my Colleagues." *Forum N.H.I.* 1, no. 1 (1994): 42–73.

Wynter, Silvia. "Unsettling the Coloniality of Being/Power/Truth/Freedom: Toward the Human, After Man, its Overrepresentation—An Argument." *CR: The New Centennial Review* 3, no. 3 (2003): 257–337.

Yampolskiy, Roman V. "Analysis of Types of Self-Improving Software." In *Artificial General Intelligence: 8th International Conference, AGI 2015, Berlin, Germany, July 22–25, 2015 Proceedings*, edited by Jordi Bieger, Ben Goertzel, and Alexy Potapov. Cham, Switzerland: Springer, 2015.

Yampolskiy, Roman V. "On the Controllability of Artificial Intelligence: An Analysis of Limitations." *Journal of Cyber Security and Mobility* 11, no. 3 (2022): 321–404.

Yonck, Richard. *Heart of the Machine: Our Future in a World of Artificial Emotional Intelligence*. New York: Arcade Publishing, 2020.

Yudkowsky, Eliezer. *Creating Friendly AI 1.0: The Analysis and Design of Benevolent Goal Architectures*. San Francisco: The Singularity Institute, 2001.

Zornad, Joseph, and Sara Reilly. *The Cinematic Superhero as Social Practice* (Cham, Switzerland: Palgrave Macmillan, 2021.

INDEX

abilisk, 20, 105

abjection, 28–30, 39, 52, 55, 59, 96

ableism, 13, 138n65

Absorbing Man, 7, 8

administrative violence, 30

Advanced Threat Containment Unit, 51–54

affective computing, 72–73

Afterlife, 44, 49, 53, 115

Agents of S.H.I.E.L.D., 8, 31, 34, 38–63, 65, 86–93, 104, 115, 121, 124, 126, 127

AIDA, xvii, 65, 86–88, 90–94

Aitken, Stuart, 29

Alien Nation, 101

androids, ix–xvii, 5, 14, 18, 20, 29, 37, 40, 65–66, 74–76, 86–90, 93–98, 101, 129–30

anthropocentrism, x, xiii, xv, 11, 12, 15, 17, 21, 75, 118

Arrival, 100

artificial general intelligence, 69–74

artificial intelligence, x, xiv, 10, 17, 20, 65–72, 75–78, 79–94, 97, 118

artificial narrow intelligence, 69

Asgard/Asgardians, xvii, 96, 106, 115–16, 120, 162n62, 163n63, 163n64, 163n66

Asimov, Isaac, 65, 73, 84, 91

Asimov's "Three Laws of Robotics," 73, 84, 91

astrobiology, 102–4

Avengers, The, 25, 32, 81, 96, 98, 107, 126, 127, 144n30

Avengers: Age of Ultron, 5, 14, 23–24, 65, 75, 79–86, 93, 118, 143n30

Avengers: Endgame, 4, 98, 125

Avengers: Infinity War, 110

Avengers Initiative, 37, 38, 48, 53

Banks, Luther, 52

Banner, Bruce, 39, 82–86, 129, 143n30

Barnes, Bucky, xvii, 126, 166n29

Barsam, Ara, xv

Barton, Clint, xii, xvii, 22, 124, 128

Battle of New York, 40, 58, 62, 81, 93, 109, 128

Bauman, Zygmunt, 122

Beal, Timothy, 26

Bernat, James, 17

biocentrism, xv, 14–17

biotechnology, x–xi, 10, 51, 94

Bishop, Kate, xi, 22, 23, 128

Blackford, Russell, 5, 6, 66, 74, 83, 98

Black Panther, xi, 142n26

Black Widow, xi, xvii, 126, 128, 143n30

Bladerunner, 83

Bostrom, Nick, 117

Braidotti, Rosi, 10, 13

Bridle, James, x, 16, 67, 69

bureaucracy, 32–38

Butler, Judith, 25

Campbell, Lincoln, 43–44

Čapek, Karel, 75, 83–84

Captain America, xi, xii, xiv, xvi, 4, 5, 10, 48, 82, 124, 133n31, 142n26, 166n29

Captain America: Civil War, 37

Captain Marvel, xi, xvii, 96, 97, 104, 107, 115

Carter, Sharon, 36

185

INDEX

Castle, Frank, 19–20, 22, 140n94

Caterpillar Program, 53–54

Chavez, America, xii

Chitauri, xvii, 57–58, 81, 96, 107, 109, 118, 128

Cho, Helen, 83, 86

classism, 138n65

Claeys, Gregory, 28, 58, 96

Cleary, P., 62–63

colonization of space, 117–18

community, xiv, xv, 13, 15, 29–30, 33, 55, 116

Coulson, Phil, 40–61, 88–89, 116, 122, 126–27, 129, 162n62

Creel, Carl, 7, 8

dangerous subject(s), 36

Danvers, Carol. *See* Captain Marvel

Daredevil, 19, 40, 55

Dartmouth Conference, 68

Day the Earth Stood Still, The, 99–100

death, 17–18, 121–23

Deathlok, xvii, 47

De Dauw, Esther, xi

Deep Blue, 69, 91

Deever, Sadie, xiii, 62–63

Defenders, The, 40

de Goede, Marieke, 33, 35

dehumanization, 14, 28, 30, 55, 58, 78–79, 90

Dick, Philip K., 74

disability studies, ix, xi

District 9, 101

Don't Look Up, 119

Dormammu, 96, 118, 124

Drax, 20

Dr. Strange, xi, xvii, 39

Echo, xii

Ellis, Matthew, xii, xiii, 51–52

Eurocentrism, 13

exceptionalism: American, xi; human, x, 15, 16, 21, 30, 31, 55, 57, 92, 104, 119, 129

exclusion, xi, 13, 29, 39, 41, 51, 58, 64, 79, 138n65

existential risks, 117–19

extraterrestrial life, 96–97

Falcon, xii, xvii, 124, 142n26

Falcon and the Winter Soldier, The, xii, 124

Fantastic Four, 7, 10, 39, 40

Feige, Kevin, 40–41

Feinberg, Joel, 16

feminism, ix, 13, 107

Fitz, Leo, 41–42, 47–48, 61, 86–88, 94, 123, 126, 130

Foster, Jane, 115–16

Foucault, Michel, 30, 50–51

Frankenstein, 64–65, 71, 73, 80, 85

FRIDAY, 79

Friendly AI, 72, 78

Fury, Nick, xvi, 5, 34, 37, 40, 48, 79, 84, 97, 106–16, 124, 126, 162n48

Gamora, 20, 126

Garner, Andrew, 53, 54, 150n146

General Talbot, 46, 48, 54

genetic engineering, xi, 10, 18, 76

G'iah, 110, 114

Gomel, Elana, 10, 14

Gonzales, Robert, 48–50

Good, Irving John, 71

Gravik, 110–16

Green Goblin, xiii

Guardians of the Galaxy, as trilogy, 55, 104

Guardians of the Galaxy, Vol. 2, 20, 26

Guardians of the Galaxy, Vol. 3, xiv, 20, 127

Gunkel, David, 66

Gutierrez, Joey, 52–53

Halberstam, Judith, 28

Hammer, Justin, 81

Hawkeye, xii, xvii, 22, 124, 128

Hayward, Tyler, 109–10

High Evolutionary, xiv

Hill, Maria, 34, 40, 58, 85, 111, 112, 126

homophobia, xii, 138n65

Hulk, xiv, xvii, 10, 39, 143n30, 163n63

human(s), ix, xiii, xiv, xvi, 24, 25, 27, 31, 57, 63, 98, 129; and alien encounters, 98–101; becoming, 6–10; being, 11–14; and bigotry, ix, 31, 33, 42, 43, 60, 101, 116, 142n26;

INDEX

and discrimination, xii, 13, 14, 30, 31, 33, 47, 54, 79, 95, 101, 114, 115, 142n26; and fear, xiii, 27, 28, 31, 33, 43, 44, 51, 56–59, 65, 66, 75, 88, 93, 97, 98, 112, 115, 118, 123, 158n119; and hatred, ix, xiii, 14, 28, 33, 43, 51, 56, 58, 60, 101, 128
Humans First Movement, 57–59
HYDRA, xvii, 36, 37, 40, 45, 50, 52, 81, 110, 157n102

index, 46–47, 54–55
Inhuman Outbreak, 50–54
Inhuman(s), xvi–xvii, 8, 10, 31, 38–63, 88, 110, 111, 115, 128
Iron Man, ix, xi, xii, xvii, 5, 7, 51, 79, 80, 124, 125. *See also* Stark, Tony
Iron Man (films), 76, 79, 81, 125–27, 131n1, 157n106
Ivanov, Anton, 56, 59, 124

Jackson, Zakiyyah, 13
Jameson, J. Jonah, xiii
JARVIS, 79, 84
Jessica Jones, 40, 55
Jiaying, 44–46, 49–50, 149n105
Johnson, Daisy, xi, xvii, 41–54, 56–59, 162n62
Joint Counter-Terrorism Task Force, 36
Joint Counter-Terrorist Center, 36

Khan, Kamala, xi, 41, 62, 63
Killian, Aldrich, 131n1
Kree, xvii, 41, 43, 96, 104–5, 107–10, 115, 126, 162n62
Kristeva, Julia, 28–29

Lady Sith, 40
Lash, 53, 54, 150n146
Lee, Stan, 7, 31, 38, 61, 131n1
life, 8–10, 17–18
life model decoy, xvii, 87, 89, 92, 126, 129
Loki, xi, 81, 82, 124, 126, 143n30
Lopez, Maya, xii
Luke Cage, 40, 55
Luper, Steven, 17

Mace, 48, 61, 88
Mackenzie, Alphonso "Mack," ix, 47, 61, 66, 121
Mantis, 20, 26
Marshall, Catherine, xiii, 101
Martin, Mike, xiv
Mar-Vell, 107–9, 115
Maximoff, Pietro, 81, 85, 126
Maximoff, Wanda, xi, 81, 85, 110, 157n108
May, Melinda (life model decoy), 129–30, 166n39
May, Melinda (human), xi, xvii, 40–42, 88, 126, 127, 150n146
May, Todd, 123–24, 127
McMahan, Jeff, 20
McSweeney, Terence, 4, 5, 61, 125, 126, 158n119
meaningful life, xv, 59, 120, 124, 125, 128, 164n3
meaning in life, 3, 91, 94, 121, 125, 127–28
Meyer, Marvin, 25
militarism, xi, 61
mind-uploading, 125
Minsky, Marvin, 68, 71
monster/monstrous, xiii, 5, 27–28, 31, 44, 46–47, 53, 64–65, 84–85, 95, 143n30
moral agency, 74–75, 85, 92, 94
moral patiency, 74–75, 90, 92
moral standing, xiv–xv, 6, 13–17, 20–21, 30, 51, 59, 75, 85, 100–103, 138n65
more-than-human beings, xiii–xvi, 5–6, 10, 14, 25–63, 66, 72, 94, 98, 101, 106, 109, 124, 129, 163n63
Ms. Marvel, xi, 41, 62, 63
Murdock, Matt, xii, 10, 19–20, 22, 55, 124
mutants, xiii, 5, 14, 27, 29, 31, 37, 39–41, 62, 95, 96, 98, 101, 113, 130, 142n26
mutations, 9–10, 43

Nadeer, Ellen, xii, xiii, 57–60, 124, 128
Nadkarni, Samira, xi, 49, 124
nanotechnology, x, 10, 18, 110, 118
Newman, Kim, 31
Nussbaum, Martha, 121–22

188 INDEX

Omohundro, Stephen, 70–71, 93
other-than-human beings, ix–xvii, 5, 6, 12,
    14, 20–26, 28–31, 33–37, 39, 45–63, 66, 70,
    74, 84–94, 98–105, 109, 114–15, 124–30

Parker, Peter, xii, xvii, 61–63, 124
Peterson, Mike, xvii, 47
Picard, Rosalind, 72
posthumanism, x–xi, xiv, 13
posthuman future, xv, 6, 11, 14, 25, 31, 35, 56,
    58, 61, 63, 92, 97, 117, 119, 126
posthuman universe, xiii, xv, 5, 7, 18, 43, 48,
    59, 66, 75, 83, 85, 90, 97, 124
Potts, Pepper, 3, 125–26
Price, Rosalind, xii, xiii, 52–53, 124
Project Cataract, 109–10
Punisher, The, 19–20, 22, 140n94

Quake. *See* Johnson, Daisy
queer studies, ix, 13
Quill, Peter, xi, xvii, 20, 26, 163n66

race/racism, xi, xii, 30, 55, 116, 124, 138n65
Radcliffe, Holden, xiii, 65, 86–92
Raina, 43–44
Rambeau, Maria, 108–10
Rambeau, Monica, xi, 110
Razer Fist, xvii
Red Skull, xii
Reinhardt, Werner, 45–46, 49, 110
reverence for life, xv–xvi, 6, 21–24, 25, 31, 47,
    55, 73, 89, 93, 100, 104, 111, 113
Rhodes, James, xvii, 51
Rieder, John, 99
Robinson, Ashley, 5
robot apocalypse, 64, 75, 86
Rocket Raccoon, 20, 127
Rodriquez, Elena, xvii, 51, 54, 88
Rogers, Steve. *See* Captain America
Romanoff, Natasha, xi, xvii, 126, 128, 143n30
Ross, Everett, 36
Ross, Thaddeus, 36–37

Sagan, Carl, 104
Scarlet Witch, xi, 81, 85, 110, 157n108
Schöfeld, Martin, xiv

Schweitzer, Albert, xv–xvi, 18, 21–24, 25, 31,
    58, 61, 73, 94, 100, 108, 128–29
Secret Invasion, 106–16
Selvig, Erik, xiii, 115–16
Sen, Amartya, 29
Sentient Weapon Observation and Response
    Division, 109–10
sexism, xii, 138n65
Shang-Chi, xi, 142n26
S.H.I.E.L.D., ix, 34–37, 40–62, 88, 110–11,
    115–16
Sibley, David, 29
Simmons, Jemma, 41–42, 61, 94, 123, 130,
    162n62
Singer, Peter, 15–16, 20
Sitwell, Jasper, 40
Skrull, xvii, 96–97, 105–18, 120, 162n48
Skye. *See* Johnson, Daisy
Sokovia Accords, 36–38, 40, 59, 81, 86
Spade, Dean, 30, 33, 34
species-being, 16, 20, 22, 40, 41, 43, 48, 67,
    74, 93, 96, 99–101, 105, 120, 122
speciesism, 15, 16
Spider-Man, xii, xvii, 61–63, 124
Stapledon, Olaf, 27
Stark, Tony, xi, xiv, xvii, 3–5, 25, 37, 51, 65,
    79–86, 93–94, 97, 118, 124–26, 129, 133n31,
    157n106, 157n110, 158n119
Star Lord, xi, xvii, 20, 26, 163n66
Strange, Stephen, xi, xvii, 39
Strategic Aerospace Biophysics and
    Exolinguistic Response, 112
Strategic Defense Initiative, 82
subject-of-a-life, x
super soldier serum, xvii, 47
synthetic biology, 10, 76

Talos, 107–15
Taylor, Affrica, 13
T'Challa, xi, 142n26
Terminator franchise, 74–77, 112
Terrigenesis, xvii, 43–45, 50, 52, 150n146
Thanos, xiv, 98, 109, 110, 116, 118, 124
Thor, xi, xii, xiv, xvii, 84, 86, 96, 107, 115, 116,
    135n8, 158n123, 162n62
transhumanism, 65, 85–87, 91, 117–19

Ultron, xvii, 23, 65, 82–86, 92–94, 113, 118
US Department of Damage Control, 61–63

Vanko, Ivan, 81
vigilantes: Avengers as, 36, 55; humans as,
    55–60, 115
Vision, 5, 15, 23–24, 36, 84–86, 94, 110, 126
Vismann, Cornelia, 35

War Machine, ix, xvii, 51
Ward, Grant, 34, 41
Watchdogs, 54–60, 112
Wells, H. G., 98–99
Whitehall, Daniel, 45–46, 49, 110
will-to-live, xvi, 21–23, 31, 61, 71, 85, 89–94,
    99–101, 104, 108, 120, 125, 128–30
Wilson, Sam, xii, xvii, 124, 142n26
Winter Soldier, xvii, 126, 166n29
Wong, xi
Woo, Jimmy, xiii, 110, 124

X-Men, 10, 31, 38–40, 43, 61, 142n26

Yampolskiy, Roman, 72
Yon-Rogg, 107–9
Yo-Yo, xvii, 51, 54, 88
Yudkowsky, Eliezer, 72

Zabo, Calvin, 45–46
Zornado, Joseph, 5

ABOUT THE AUTHOR

James A. Tyner is professor of geography at Kent State University and a Fellow of the American Association of Geographers. He received his PhD in Geography from the University of Southern California. Tyner is the author of more than twenty-five books, including *War, Violence, and Population*, which received the AAG Meridian Book Award for Outstanding Scholarly Contribution to Geography. In addition, Tyner is the recipient of the AAG Glenda Laws Award for outstanding contributions to geographic research on social issues. His research coalesces around violence, genocide, militarism, and political economy.